Twisted
Reckonings

Philip DeLizio

First paperback edition June 2024

www.delizioauthor.com

Published by Hemingway Publishers

Cover design by Hemingway Publishers

**HEMINGWAY
PUBLISHERS**

ABOUT THE AUTHOR

Dr. Philip DeLizio, Ed. D. is a retired schoolteacher. Upon his retirement, he began to write crime/mystery books based in the Caribbean. He currently resides in Maryland and travels to the Caribbean several times a year.

Books

Young Adult Inspirational

Rekindled Faith

Emma's Dilemma

A Light on the Horizon

Darkness to Light

A Letter of Hope

Crossroads

Paul Phillips Mysteries

Twisted Reckonings

Shadows in the Jungle

The Missing Piece

The Cricketer's Conspiracy

Shadows of the Past

DEDICATION

For Anthony, My Beloved Son

Retired detective Paul Phillips wants nothing more than a quiet life in the Caribbean, away from the dark mysteries of his past. But when wealthy businessman Ray Thompson is found dead under suspicious circumstances, Paul reluctantly agrees to assist ambitious rookie detective Zoe Walker with the high-profile case. As Paul and Zoe delve into Ray's history, they discover the victim has just as many secrets as enemies. Their investigations stir up tangled threads connecting Ray's demise to unsolved crimes from Paul's previous cases. Meanwhile, Zoe pushes boundaries in her drive to close her first major investigation, testing Paul's patience and expertise. Clues led them to several potential suspects, each with motives and secrets of their own. However, as Paul and Zoe work to untangle the web of lies and deceit surrounding Ray, someone is determined to cover their tracks. Just when they crack the case wide open, a shocking discovery reveals the killer has been one step ahead all along. With lives now at stake, can Paul and Zoe overcome the tensions between them to solve the mystery before another dies? To prevail, they must venture into the depths of Paul's own haunted past and uncover the twisted truths at the core of this troubled present.

TABLE OF CONTENTS

CHAPTER 1

A PAST NOT FORGOTTEN

Paul Phillips is a retired detective who has seen it all. His years on the force have left him with a cynical outlook on life but also a deep-seated sense of justice and duty. He is a man of few words, preferring to observe and gather information before forming opinions. He is patient and methodical, traits that served him well in his former career. However, he can also be gruff and impatient when dealing with those he perceives as incompetent or untrustworthy. Despite his tough exterior, Paul cares deeply about the people around him, especially those who have suffered at the hands of criminals. He is not one to seek recognition or praise for his actions but rather finds satisfaction in knowing that he's made a difference. As a retired detective living on the Caribbean Island of St. Anne, Paul enjoys spending his days relaxing on the beach, sipping rum, and fishing.

St. Anne is a Caribbean gem that holds a special place in Paul's heart. This island is a vibrant tapestry of natural beauty and intriguing history. For Paul, what makes it truly captivating is its rich French-infused heritage. St. Anne is blessed with stunning landscapes that seem straight out of a postcard. From the pristine white sand beaches, gently lapped by the azure waters of the Caribbean Sea, to the lush rainforests teeming with exotic flora and wildlife, it is a paradise for the senses. But what really sets St. Anne apart is its history, steeped in French influence. The island was once a French colony, and remnants of that heritage are visible everywhere. The charming architecture, the quaint fishing villages, and the delicious culinary delights all bear the mark of French culture.

One cannot help but fall in love with the island's unique blend of Caribbean charm and French sophistication. The friendly locals, known for their warm hospitality, add to the allure of this remarkable place. Whether it is exploring the vibrant local markets, indulging in authentic French-Caribbean cuisine, or simply strolling along the waterfront, there is always something new to discover. For Paul, St. Anne is not just a beautiful island; it is a refuge. It allows him to escape the shadows of his past and immerse himself in the tranquility of its surroundings. The soothing sound of the waves and the breathtaking sunsets serve as a constant reminder that there is still beauty and peace to be found in this troubled world. It is a place where he can find peace and perhaps even redemption.

The soothing crash of waves gently woke Paul from his midday slumber, a light sheen of sweat bathing his bare chest in the tropical heat. Shading his eyes, he gazed out at the azure waters stretching to meet the cloudless sky, remnants of yesterday's storm leaving the azure sea serenely flat.

Another day in paradise, or so it should have been.

But as Paul rolled over, fingers of tension kneaded the back of his neck where memories often lurked. Five years away from that life, yet its shadows still haunted him, sins of the past seeping into this haven that was meant to drown them. He had hoped the simple tranquility of island life could finally quiet the ghosts. Foolish thought, as time had proven.

Rising with a sigh, Paul pulled on a light linen shirt and walked barefoot along the beach, toes curling in the fine, warm sand. Gulls cried raucously overhead, their cries strangely melancholic against the backdrop of lapping waves. He drifted aimlessly, hoping the familiar motions of tide and shore could scrub his mind blank, if only for a moment.

It was then a glint in the sand caught his eye, a curved sliver of aqua-hued glass, smoothed by years in the sea. Picking it up, Paul turned the makeshift worry stone between calloused fingers, his mind as restless as the rippling tide. The past refused to relinquish its hold, to grant him the simple respite he craved, no matter how far he fled, and today, it seemed determined to haunt him once more.

Paul returned to his bungalow and sat in his beachfront hammock, watching the gentle waves lap against the powdery white sand. He took a sip of his favorite rum and coconut juice, savoring the smooth, sweet taste on his tongue. This was his life now: retired detective, living on the beautiful island of St. Anne. He had earned it, damn it, and he was going to enjoy every second of it.

Standing at 6'1", his toned build speaks of his former athleticism. His hair, now silvered with age, is neatly trimmed and combed back. His eyes are a piercing blue, often holding a hint of weariness from years of sleepless nights spent chasing criminals. He has a well-groomed beard that is starting to turn grey as well. Paul's skin has developed a leathery texture from years spent in the sun, giving him a weathered appearance that is nonetheless distinguished. He prefers casual attire such as shorts, t-shirts, and sandals, accented with a worn leather belt and a pocket watch.

Then it happened: a knock sounded at the back door, the door nearest the driveway. The door his friends never used. Paul made his way to the door, dreading what, or who, might await on the other side. He flung it open to reveal a young woman clutching a worn backpack.

"Detective Phillips?" she asked, hope and apprehension worn on her face without even a hint of trying to hide her excitement from Paul. "I'm Detective Zoe Walker from the St. Anne Police Department. I was told you used to work cases here before retiring, and I could really use your help with an investigation."

Paul tensed, memories surging forth unbidden.

Zoe, at 5'5", has a slender yet curvaceous figure. Her long raven hair cascades down past her shoulders, often falling over her face when she is deep in thought or concentrating on a case. Her green eyes are sharp and observant, always taking in every detail around her. Despite the constant heat of the Caribbean, she prefers to wear comfortable yet stylish clothing, often seen in a black blazer, fitted pencil skirt, and a crisp white button-up shirt. To complete the look, she wears a pair of black flats and carries a sleek black leather purse. Her almond-shaped eyes were piercing, and her features were a mix of French and African descent, reflecting the island's rich cultural heritage.

Zoe is highly intelligent, analytical, and dedicated to her work. She has an intense sense of justice and always strives to uphold the law. Her persistence and attention to detail make her an excellent detective, but she can also be quite impatient and impulsive at times. She values her independence and does not like to rely on others, which sometimes makes her come across as aloof or cold. Despite this, she has a soft spot for those who are truly in need and is always willing to lend a helping hand. Her emotions can sometimes get the better of her, especially when she becomes too invested in a case.

Born and raised in St. Anne, Zoe spent much of her time outdoors, swimming in the ocean, exploring hidden coves, and hiking through the dense rainforests. She immersed herself in the culture of the Island: colorful festivals, lively music, and delicious Caribbean cuisine. She has fond memories of dancing to calypso and reggae music during local festivals when she was younger. But it was not just all fun and games. St. Anne also had its fair share of hardships and challenges. Poverty and crime were present, which was the driving force in her joining the police department. She wanted to make a positive difference in the community. Growing up in St. Anne shaped Zoe into the person she is today. It taught her resilience, compassion, and the importance of fighting for justice. It's truly a place she is proud to call home.

When it comes to personal relationships, Zoe values the people in her life deeply. She has a small yet close-knit group of friends who have been there for her through thick and thin. It is important to Zoe to have that circle of trust and camaraderie, especially in a profession like hers where the work can be intense and emotionally challenging.

As for love interests, Zoe's focus has mostly been on her career as a detective. She has a lot of ambition and drive to make a difference, and that has taken precedence in her life. However, she secretly admits to herself she believes in love and companionship, and if the right person comes along, she is open to exploring that side of life. But for now, her focus remains on honing her detective skills and solving cases. If she has learned anything, she has learned that it's important to strike a balance between work and personal life. It is something she is still figuring out, but she believes in finding someone who understands and supports her passion for justice while also being there for her as a partner. Love will come when the time is right, and until then, Zoe is content with building her career and nurturing the relationships she already has.

At the sight of Zoe, Paul's memories returned: Five years ago, a wealthy businessman was found dead in his office, with no sign of intrusion or struggle. As lead detective, Paul had meticulously picked apart every minuscule clue, following a twisting web of secrets and deceit. It led him to make a harrowing discovery, one linking the businessman to far darker evils. The killer sensed his imminent capture and struck, taking Paul's partner hostage to lure him into a trap. He arrived only to find his friend bleeding out, the killer's sinister laugh echoing through the warehouse. In the ensuing confrontation, three shots rang out, then silence.

When backup arrived, they found Paul cradling his dying friend amid a pool of blood, the killer's body several feet away, and a gun that had fallen from his lifeless hand. The trauma was too much. Though he bagged the perp, Paul was never the same. Nightmares filled his waking hours until he decided a quiet retirement was the only way to find peace.

Snapping back, Paul frowned at Zoe. "I am sorry, detective, but I'm long past that life." But then asked, "What is the case you are working on?"

Zoe took a deep breath. "A businessman named Ray Thompson was found dead in his home two days ago. It's my first big case, but there are already questions about the investigation. The chief thought having your expert eye could help."

Paul sighed, walking towards the beach, tracing patterns in the sand as the waves lapped reminders of his pledge to leave this all behind. Looking out at the ocean, he felt a mixture of annoyance and dread in his gut. He knew the score. He knew what taking this case would mean. He would have to abandon his peaceful, quiet life on the beach, delve back into the dark underbelly of the island, and face the dangers that came with being a detective. But he was a detective. It was in his blood, much to his annoyance. And deep down, he knew he could not turn his back on the case. Not when it meant bringing justice to the innocent.

He had to ask, "Why come to me?"

Hope flickered in Zoe's green eyes. "The coroner found lacerations indicating the killing was personal, and Thompson had a shady past. His records show your name from a cold case that was never solved. I think it's linked that you're the only one who can help me get justice for both."

Straightening, Paul gazed toward the sea, torn between duty and peace. This place was now his haven, yet it had not saved him from ghosts' past. And if closing the cold case could exorcise them at last, He turned back to Zoe, resignation in his tone.

"Very well, detective. Show me what you have so far. But I make no promises beyond consulting—I'm long retired from this line of work."

Zoe offered a hesitant smile. "Thank you, sir. Any

insight could help. This way to my vehicle. If you will come with me?" With a final glance at the tranquil shore, Paul steeled himself and followed Zoe inland, slipping back into a life he had sworn to leave behind. The drive to the crime scene was quiet, neither speaking, each waiting for the other.

After 25 minutes of silence, they finally arrived at the sprawling beachfront mansion. Approaching, one first notices the grandeur. The estate is like something from a postcard, with its white walls and vibrant gardens. The palm trees were swaying gently in the breeze, casting dappled shadows on the pristine lawn. The home itself is an architectural marvel, a blend of modern luxury and classic elegance.

Entering the home, it is hard not to notice the expansive floor-to-ceiling windows, which allow for breathtaking views of the turquoise waters just beyond. The interior is equally impressive, with spacious rooms adorned with opulent furnishings and tasteful artwork.

With Zoe taking the lead, Paul followed, breezing through an impressive foyer, now marred by grim investigation. In the magnolia-walled study, the coroner stood over a sheet-draped form on the Persian rug.

"We are almost done here," said the coroner, acknowledging Zoe's arrival, then departed silently.

Zoe, looking at the lifeless body lying on the floor, admitted to herself, not wanting her partner to know, hit her in a way she knew she would not soon forget. The crime scene, she thought, was somber, and the atmosphere was heavy with a sense of loss and mystery. However, as a detective, her focus was immediately drawn to the details, trying to piece together the puzzle and uncover the truth behind the crime.

Taking notes, she noticed the victim lying motionless on the floor. It was a chilling sight, seeing a life extinguished so suddenly. His body was sprawled out, surrounded by a pool of blood, adding to the grimness of the scene.

The silence was deafening, broken only by the low hum of conversation as investigators moved around, meticulously collecting evidence. As she approached the body, careful not to disturb anything, she noticed the signs of struggle. It was clear this was not just a random act of violence but an intentional act to end someone's life.

Crouching, Paul slowly folded back the sheet to

reveal gaunt features frozen in a rictus of shock and pain. He donned gloves and meticulously examined the body, years of habit rising unbidden. Two deep lacerations to the torso, defensive wounds, and signs of struggle.

The murder victim, Ray Thompson, was brutally stabbed to death. The scene was one of immense violence and tragedy. The body lay sprawled on the floor, lifeless, a grim reminder of the darkness that lurks in the world. Thompson's wounds were deep and numerous, evidence of a frenzied attack. The sharp blade of the weapon had left a trail of devastation upon his body, each wound telling a story of pain and desperation. The air was heavy with the metallic scent of blood, lingering as a reminder of the gruesome act.

The sight was not for the faint of heart, as evidenced by Zoe's face. It was the kind that haunts the memories of those who bear witness to such horrors. As a seasoned detective, Phillips had seen his fair share of gruesome scenes, but this one was particularly chilling. The intensity of the violence is a stark reminder of the darkness that exists within humanity.

Scanning the opulent yet impersonal room offered no witnesses or valuables as motives. He gingerly shifted the corpse, noting discoloration along the flank. Lividity suggests death at least 12 hours prior to discovery in this location. But what happened in between?

The victim, Ray Thompson, 45 years old, had a dark complexion, common among people from the Caribbean Islands. He was tall, easily measuring 6 feet, with a muscular build. His hair was black and curly, styled in a neat haircut. He was dressed in an obviously expensive tailored suit, which emphasized his athletic build. He was also wearing a gold chain around his neck and rings on his fingers.

Robbery, Paul thought, was not a motive.

Rising, Paul's gaze swept over blood-spattered furniture and tome-lined walls, soaking in every nuance with a practiced eye the others lacked. Paul thought the room was quite a unique space, blending comfort and functionality. It is hard not to notice the expansive windows that offer panoramic views of the ocean. Natural light filters in, creating a serene and calm ambiance despite the murder victim on the floor. Irony, Paul thought. The walls were adorned with elegant bookshelves, filled with a vast collection of literature on various subjects.

In the center of the room is a large mahogany desk, meticulously organized with files, folders, and a computer. Paul made a mental note to search the computer files for possible clues. The desk is positioned to maximize the view of the ocean as if to bring inspiration to whoever sits there. A comfortable leather chair sits behind the desk, perfectly suited for long hours of work. Adjacent to the desk, a seating area with plush sofas and armchairs is arranged around a low coffee table. A small bar cart tucked away in one corner, stocked with the finest spirits to offer a touch of indulgence.

He crouched to inspect a scuff on polished wood, sniffed a trace of perfume, and tilted his head at the faintest disturbance in the dust on a display case across the room.

"Detective Walker." Paul started, "I suggest we look deeper into our victim's life and those who knew him best. Some secrets still hide in plain sight here."

Zoe stared, impressed by his deductions. "Lead the way, sir. With your expertise, we can solve this case together."

Returning to the beach house as night fell, Paul filled his lungs with tangy sea air, hoping its familiar balm could scrub the day's grimness away. But Zoe followed him inside, green eyes full of earnestness.

"I know you said you are retired," she began, "But we are losing valuable time. This could be connected to an old case. Does that matter to you at all?"

Paul sighed, his gaze drawn to the crashing waves, now dark silhouettes against the inky sky. "This place was meant to be my life raft away from that world. I've found peace here."

She shook her head stubbornly. "Or maybe you are still running from the past. With your expertise, we could crack this case and get justice for both victims. Is not laying the ghosts to rest worth it?"

His jaw tensed. "I made my choice long ago. Now it is time for you to leave, detective, before the night sets in. This is as far as I go."

A flash of anger broke through Zoe's composure; then, her shoulders slumped in defeat. "Thank you for your time, sir. I appreciate you humoring me this far. Good evening."

Paul watched her small figure retreat into darkness,

guilt prickling his skin. Did she have a point about the solace of resolution versus escape? No, he had fled that life for good reason. This was over. Or so he told himself as waves crashed a restless refrain.

That evening, Paul walked the beach, waves sighing their eternal lament. Did he genuinely want this bucolic half-life, or was he still running?

Another memory: Five years ago. It was another late night at the precinct, immersed in case files, seeking any thread to break the whole case. Paul had pored over photos of their latest crime scene, rubbing tired eyes. Then, a scrawled note caught his gaze, an unfamiliar symbol in the margin. This memory sparked something, and hours later he pulled a cold case, noticing the same marking. His mind raced, piecing disparate clues together. This symbol interconnected them all; they were dealing with a serial killer. But who was pulling the strings? Bursting into his captain's office at dawn, Paul's revelation was met with stunned silence. Then the call came, another body was found, the same symbol left in blood nearby. They raced to the scene, but it was too late.

Shaking free of the memory, Paul stared into roiling

waves as if they held answers. Had he solved that last case, would things have ended differently? Or was escaping truly the only way to survive? With a sigh, he turned from the sea, its lulling rhythm unable to calm his restless soul that night. The storm within refused to abate, and for the first time, Paul wondered if facing the tempest might finally still its rage.

In the morning, after a less-than-fitful sleep, Paul strode into the sparsely decorated St. Anne PD office. Zoe glanced up in surprise from her desk.

"Detective. I underestimated the gravity of this case. I will consult on one condition."

Zoe set down her pen expectantly. "Name it."

"You accept my experience without question. I lead the investigation my way." She considered his stern gaze. While used to being in charge, something in his eyes said this was non-negotiable.

"Very well, sir. On one condition of my own, you keep me informed every step and help me learn."

Paul extended his hand. "Then we have an accord, detective. When do we start?"

Zoe grinned, shaking firmly. "Right away, sir. I pulled Thompson's records; there's more to his past than meets the eye. Thompson was ambitious and ruthless, known for his shady business dealings. From what I have found so far, he has quite an extensive file; he was manipulative and intimidating, and many people feared him."

Zoe continued, "However, he was also known to be generous with his wealth, especially with his mistresses and close associates. He apparently led an extravagant lifestyle. From some of the transcripts I have read so far, he was quick to anger and could be violent when provoked."

Rising, Paul nodded. "Let us delve deeper then. A secret this man held could be the key to unraveling this."

With the storm clouds of his past lifted, if only temporarily, Paul followed Zoe to peruse the files. Their partnership sparked, and the hunt renewed. The game, it seemed, was afoot. Sitting across from Zoe in the break room, Paul blew steam from his coffee, regarding her keenly.

"Tell me about your career so far, detective."

She met his gaze steadily. "Grew up here, went to the academy, top of my class. Been with the St. Anne

Department for two years, working on mostly smaller cases to cut my teeth. This is the first real chance to prove myself." Her passion was evident, yet Paul sensed her rawness. Solving murders required experience she lacked. But he needed her resources and youthful perspective, however grudgingly.

"Your ambitions are admirable. But this investigation will test you. Are you truly prepared to face what we may uncover?"

Zoe set her jaw. "I know I have more to learn, which is why I asked for your help. You will find I am a quick study, and I want justice just as much as you."

Paul exhaled quietly. Her fire could prove an asset if properly guided or a detriment if left unchecked. Only time would tell which path lay before them. Rising, he said, "Come, detective. Let's review those files and see if any hold promise." His mind churned with uncertainties about working as a team, yet the chase was afoot once more. For better or worse, their fates were now intertwined.

CHAPTER 2
SECRETS IN DEATH

The sterile lights of the morgue cast an eerie glow on Thompson's lifeless body as Paul examined it while Zoe read the coroner's report aloud.

"The coroner determined that Thompson's cause of death was multiple stab wounds inflicted with a sharp and pointed weapon. The wounds were deep and precise, indicating a deliberate and targeted attack. The coroner noted that the pattern of the wounds suggested a frenzied yet methodical assault, leaving little doubt about the intensity and brutality of the attack."

Zoe had to pause her reading. Being at the morgue with the murder victim's body and reading the coroner's report was intense. She felt a mixture of emotions: a sense of sadness and empathy, realizing that someone had lost their life in such a brutal way. But alongside that, there was a strong drive to find justice for the victim, a determination to uncover the truth and bring the culprits to account. It is moments like these that remind Zoe of the importance of her work and the responsibility she has to seek justice for those who can no longer speak for themselves.

Zoe continued, "The coroner's analysis revealed that Thompson had defensive wounds on his hands and arms, indicating that he had fought to defend himself against his assailant. The presence of these defensive wounds suggests a struggle and a desperate fight for survival. Additionally, toxicology reports indicated the absence of any drugs or substances in Thompson's system, suggesting that he was not under the influence at the time of the attack. This finding supports the idea that Thompson was fully aware of the danger he faced and fought before succumbing to his injuries."

Paul, taking in what Zoe had said, responded, "I believe if we retrace Thompson's last known steps, which is where we will find what happened to him and why." Paul gazed thoughtfully at the ravaged corpse. "And maybe some clues to who wanted him silenced for good." Paul leaned in closer, tracing a faded bruise near the elbow. "See this discoloration? The shape is indistinct, but these could be fingerprints from a scuffle. Someone tried to subdue

Thompson long enough to kill him."

Zoe scribbled Paul's observations in her notepad, struggling to keep up. His eyes scanned the corpse like a predator dissecting its prey. There was an intensity to his examination that hinted at his success as a detective but also darkness she found unsettling. Carefully cataloging every discernible mark and injury, Paul spoke almost to himself as he worked. "There, under the fingernails... some sort of fibers. I want those analyzed for trace evidence."

"And the bruising on the wrists?"

"Two assailants working in tandem is my guess."

Zoe frowned, overwhelmed by the wealth of clues Paul uncovered with just a glance. It was clear his experience far surpassed her own. For the briefest moment, she questioned if she was truly ready to take on such a complex investigation. But determination overcame her doubts. This was her chance to learn from the best and finally make her mark. Instead of dwelling on her shortcomings, she views it as an opportunity to learn and improve, appreciating the fresh perspective and insights that a senior detective can bring to the table. She acknowledges to herself that all of them have different experiences and areas of expertise. She had long ago decided to set aside any personal ego and work collaboratively to solve the case. Ultimately, it's about the pursuit of justice, not personal glory.

Paul gripped the victim's wrist, rotating it to examine a tattoo partially obscured by the bruising. A series of numerals and symbols held meaning for the former detective. "This tattoo... I have seen it before. Back during my time on the force," Paul said pensively. "Look into Thompson's history. See if you find any connections to organized crime or the local underworld. That marking was used by a local gang I busted years ago."

Zoe made a note to research Thompson's background for ties to criminal elements. "You think his death was somehow gang-related?"

"I'm not sure yet. But this tattoo suggests there is more to our victim than meets the eye. Whatever enemies he made or business he did in those circles could be integral to his demise." Paul released the wrist and stepped back.

"I will get started pulling his records now," Zoe said, locking the new clues away in her mental case file. This cryptic mark on Thompson's body proved Paul's savvy instincts were already steering them somewhere the local police had overlooked.

Leaving the coroner to complete her job, Paul asked Zoe to go to the file room for further research. Paul decided to head to Thompson's office. He told Zoe he had a hunch that there might be some important documents or records that could provide insight into his activities and potential motives behind his murder. Paul was a firm believer that a person's workspace can often reveal much about their character and the secrets they might be hiding.

"I will meet you at the office in a couple of hours," Paul said to Zoe. After Paul asked Zoe to go to the file room while he headed to the victim's office, she was initially a bit frustrated. She wanted to be there with Paul, checking out the scene and gathering evidence. But she also understood that he wanted her to start looking into Thompson's background and past cases, so Zoe saw it as an opportunity to prove herself and show him what she was capable of. She knew he had his reasons for wanting to check out the victim's office alone. Reluctantly, she took a deep breath, put on her detective hat, and headed to the file room, ready to dig into those records and find some leads.

Zoe met Paul a few hours later with a file folder containing everything she had uncovered about Thompson. "As I suspected, he had his hands in many dubious pies. Racketeering, money laundering, narcotics, you name it, he profits from it. That kind of enterprise leaves a lot of bitter rivals."

Paul glanced through the documents chronicling the businessman's criminal exploits. "I see at least five names that stand out as having potential motives," he said. "Any of them could have wanted Thompson dead: Carlos Diaz, Simon Ames, Alexa Chavez, Frankie Morales, and Tyler Greene."

Paul was familiar with all five names from his time on the island. He wished he was not. He started with Carlos Diaz. "Carlos Diaz, the head of a competing crime family, had a contentious relationship with Ray Thompson. Their connection was layered with an intricate web of power struggles, illicit dealings, and personal vendettas. Carlos and the victim were not strangers to each other. Their relationship was one of mutual enmity, fueled by clashing ambitions and conflicting interests. As the head of a notorious crime family, Carlos saw Ray Thompson's legitimate business empire as a threat to his criminal operations. Ray's influence and connections presented obstacles to Carlos's power and control. This, in turn, led to clashes and confrontations as they fought for dominance in their respective territories. Their rivalry extended beyond business matters. Carlos held a deep-seated grudge against Ray, stemming from a past incident where Ray betrayed Carlos's trust and jeopardized his criminal activities. This betrayal, in Carlos's eyes, resulted in significant losses and damage to his reputation, creating enduring resentment.

"Simon Ames. A disgruntled investor. Thompson and Ames had quite a contentious relationship. Thompson swindled millions from Ames in a failed business venture. Ames had put his trust in Thompson, believing he was making a sound investment. But as it turned out, Thompson's promises were nothing more than empty words. Their partnership quickly soured, leaving Ames feeling betrayed and seeking revenge. Their animosity ran deep, providing Ames with a strong motive for wanting Thompson dead.

"Alexa Chavez. Thompson's former lieutenant and former mistress. She was a complex character who alleged sexual misconduct when she left." Now, thinking to himself, Alexa was an attractive woman in her late thirties. She had sleek, dark hair that fell just below her shoulders. Her piercing green eyes held a mix of intelligence and determination, reflecting her sharp mind and ambitious nature. Alexa possessed a strong, confident presence. Standing at an average height with a lean, athletic build. Her fashion sense leaned towards sophisticated power dressing, favoring tailored suits and statement accessories. Now continuing with Zoe, "Though her appearance may have suggested a woman who had it all together, her history with Thompson was marked by allegations of sexual misconduct, revealing a darker side to their relationship.

"Frankie Morales, a local politician with rumored ties to Thompson's operation. Frankie was quite an enigmatic figure. Many whispered that he was a crooked politician, but like many rumors, the truth is often shrouded in shadows. Morales had a certain charm and charisma that drew people in, making him a popular figure in the community. He had a well-groomed appearance, with salt-and-pepper hair and a neatly trimmed beard that added a touch of sophistication to his image. Morales had a way with words, effortlessly navigating political circles and making alliances. While there were allegations of corruption and illicit activities tied to Thompson's operation, whether Morales was deeply involved in such practices remained unclear." Unbeknownst to either Paul or Zoe, as the investigation began to unfold, it would become evident that Morales had some connections to Thompson's web of secrets.

"Finally, Tyler Greene, Assistant District Attorney, who at one time was pushing to indict Thompson, then suddenly backed off the case. Greene was a sharp, ambitious individual who possessed a certain air of authority. He had a commanding presence and piercing grey eyes that held an intensity that mirrored his drive for justice. As for his sudden backing off from indicting Thompson, well, that's a matter of speculation and intrigue. Rumors circulated that Greene had received a considerable amount of pressure, both internal and external, to drop the case against Thompson.

"Some whispered of political connections, insinuating that higher powers wished to protect Thompson for their own gain. Others suggested there might have been compromising evidence against Greene, perhaps leading to a compromising situation that undermined his ability to prosecute the case properly. Whatever the true reason may be, Greene's decision to step back left many questioning his integrity and wondering how deep the roots of corruption sprawled within the system.

"All will require questioning. However, one of these stood to gain the most, with Thompson removed permanently. Our investigation begins with them."

Paul turned to Zoe. "We have our list of suspects. Now comes finding the killer hiding among them."

With Paul leaving to clear his head and get some fresh air, Zoe reviewed case files long into the night, rubbing her tired eyes. This was unlike any case she had worked on before, and the endless documents were starting to blur.

Reviewing case files into the night, Zoe thought. It is a bittersweet ritual. On the one hand, poring over those documents gave her a sense of purpose, a hope of unraveling the tangled web of clues. It is a reminder of the type of detective she wants to be. The one who could piece together disparate fragments of evidence to solve the most elusive cases.

But on the other hand, it also dredges up memories, both good and bad of the cases that she was given as a rookie. The menial cases no one else wanted. The long nights can be mentally and emotionally draining, but they're a necessary part of the process. A necessary evil, the path she must walk to find the truth, even if it means facing demons from my past.

Several hours later, longer than Zoe liked, Paul returned. "Still at it, I see," said Paul, entering with fresh coffee.

She accepted the cup gratefully. "I didn't think you were coming back," she said smartly. "Where did you go?"

His reply was surprising, "I took a moment to clear my head and gather some additional information. I went to the local library and used their Internet to delve into the archives, hoping to uncover any hidden connections or overlooked details. It's amazing what you can find buried within the pages of history. While I didn't stumble upon any groundbreaking revelations, I did manage to uncover a few intriguing leads. Sometimes, the key to solving a case lies in the overlooked details, those bits of information that others may have dismissed as insignificant. Now, whether these leads will ultimately prove fruitful or turn out to be dead ends remains to be seen, but it's all part of the puzzle, isn't it? The important thing is to keep digging, to keep piecing together the fragments until the truth reveals itself."

Zoe was not surprised to hear Paul went to the library, even if he was a bit old-school. Paul's meticulous nature and determination would drive him to uncover the truth no matter where he would have to go. But she still had to smile.

"There is just so much here," Zoe exclaimed. "All these strangers we are supposed to understand well enough to know which one killed Thompson."

"The details can feel overwhelming". Paul leaned against the edge of her desk, "But you are handling it well for your first time out."

Zoe sighed. "I'm not so sure. What if I miss something important? We could arrest the wrong person."

"Doubting yourself means you care about getting it right." He offered an encouraging smile. "But you need rest to keep a clear head. No one expects you to solve it overnight." His counsel calmed her nerves. Paul was right.

She was trying too hard to do everything alone.

"Thank you. I think I just needed to hear that."

"We all have moments of uncertainty. That is why we have partners to lean on." Paul headed for the door. "Now go home and get some sleep. The case will still be here in the morning. See you tomorrow."

Once home, Paul decided to unwind and take his mind off the day's events. He thought about reading but decided not. Instead, he poured himself a glass of his favorite Island Rum and sat in his favorite chair, listening to calming instrumental music. But he was being pulled to a stack of old case files, so much for relaxation. While reviewing the files, shadows from his past resurfaced. Again. Another gang tattoo like Thompson's stopped him cold.

He was back in that dark alley, gun drawn as the suspect fled. "Police! Stop!" But the man whirled and fired. Paul dove for cover as the bullet grazed his arm. In the ensuing struggle, his gun went off, and a shot echoed through the narrow space. When the smoke cleared, the man lay dead at his feet. The shooting was later deemed justifiable by his department, but Paul still saw the vacant eyes accusing him of his nightmares. And the gang members he put away vowed revenge.

As Paul readied for bed, he recalled opening his front door to a hooded figure, then searing pain as a blade sliced his side. He shot on instinct. Only later did forensics confirm his attacker's gang ties. The narrow escape was Paul's breaking point, leaving him shaken. Maybe that's why this case felt so personal. Its tangled threads wove back to his last unsolved case when someone ensured the gang's threats were not empty. Paul was determined to follow these connections to their end, wherever they led. Now, to try and get some sleep.

He woke early the following day to the view of the Caribbean Sea. His view is truly breathtaking and has a profound impact on his mood. The sight of the sparkling blue waters and the gentle sway of the palm trees fills him with a much-needed sense of tranquility and peace. To Paul, it is nature's own personal reminder to take a moment to appreciate the beauty around us. Seeing the Caribbean Sea first thing in the morning brings him a sense of optimism and motivation. It is also a reminder of the vast possibilities that lie ahead, both in Paul's personal life and in his pursuit of justice as a detective.

The sea's endless expanse is a reminder that challenges can be overcome and mysteries can be unraveled. There is something almost magical about the coastal air and the soothing sound of the waves crashing against the shore. It calms his mind and prepares him for the day's investigations, helping Paul approach the day with a clear and focused mindset.

Whenever he works on a tough case, along with the calming view of the water, he always tries to make time for breakfast. He believes it is essential to fuel up and keep his energy levels high, especially when dealing with a challenging investigation. A lesson he learned long ago. Some habits never die. Usually, Paul opts for something quick and nutritious; his go-to is a bowl of oatmeal with fresh fruits or a protein-packed smoothie. Paul is a firm believer that a good breakfast sets the tone for the rest of the day and helps him stay focused on solving the case.

Just after finishing breakfast, Zoe arrived at Paul's and entered his office. Paul's 'office,' if you can call it an office, is tucked away in the rear of his bungalow. It is a relatively small and cramped space, but it serves its purpose.

The walls are adorned with a few old crime-solving mementos and a map of the surrounding area. His desk is cluttered with old files, case notes, and a collection of mystery novels that he turns to for some respite. It certainly is not the most glamorous of workspaces, but it's where he can analyze clues and piece together the puzzle of the cases that come his way.

Zoe found him lost in thought, miles from the present. "Paul? You okay?"

He blinked as if waking. "Fine. Just going over leads."

Zoe was not convinced. His eyes held a haunted cast she had noticed before. "You seem distracted. Is it something to do with your past career?"

Paul shifted papers, being evasive. "Old cases sometimes come back around, that is all."

"You know you can talk to me, right?" She spoke gently, not wanting to pry.

"There is nothing to discuss." His gruff tone made it clear the subject was closed. Zoe knew better than to push, sensing his demons lingering close to the surface.

"Well, if you change your mind..." Paul's face softened in appreciation, though his gaze remained troubled. For now, old wounds were not ready to be probed. But having an ally helped, knowing he was not alone in facing this complex case's dark threads to the past.

"Let us go," Paul said almost too abruptly. But Zoe understood.

Paul and Zoe arrived at Ray Thompson's beachside home to search for clues. Before the search began, Paul explained to Zoe that looking for clues, or possible clues, was an art. An art he had perfected years ago. "First, one must take a moment to assess the room and its surroundings, allowing the details to unfold before their eyes. Pay close attention to any potential points of interest, a misplaced object, unusual marks, or anything that strikes you as out of place. Always start with a systematic approach, starting from one corner and working your way around the room. Examine every nook and cranny, paying attention to even the smallest of details. Being thorough is paramount because clues can often hide in the most unexpected places."

Zoe listened in awe. Paul continued, "Use all your senses, relying not only on your eyes but also on touch and smell. Run your fingers along surfaces, feeling for hidden compartments or irregularities, and even catch a faint scent that could lead you closer to the truth."

Of course, Paul explained with some pride that experience and intuition play a significant role in the process. "Trust your instincts, following the threads that connect the clues like a skilled weaver unraveling a tapestry. It is a meticulous process, one that requires careful attention and a discerning eye. More importantly, searching for clues is not just about finding physical evidence. It's about understanding the story the room tells and how all the pieces fit together. It's an intricate dance of deduction and observation, and with a bit of luck, the truth will reveal itself."

After receiving this advice from Paul, Zoe felt gratitude and excitement. Gratitude because she appreciated the willingness of Paul to share his knowledge and experience with her, and excitement because it meant she could learn and grow as a detective. She found it reassuring to have someone more experienced guiding her through this complicated case and giving her valuable insights. She admitted to herself the advice boosted her confidence and gave her much-needed motivation. However, she thought it better not to share her feelings with Paul. Zoe took Paul's advice to heart.

Soon after Paul's little speech, she came across a false bottom in a desk drawer. "Look at this," she glowed. Paul joined her as she was sliding out ledgers and documents that were crammed inside.

"His public dealings do not tell the real story, do they?" He opened the first ledger. Entries detailing large sums of money and cryptic notes hinted at money laundering on a massive scale.

Zoe opened a folder packed with surveillance photos. "He was also working as an information broker of sorts. Blackmail material on many powerful people."

"So, the respectable businessman act was just a front," Paul mused. "No wonder he had so many dangerous enemies. He was lucky to have survived this long in his line of work." This discovery showed just how deeply Thompson was involved in the underworld. It gave their suspects the motive to silence him forever to protect even bigger secrets buried in these damning files. The victim's life was far murkier than initially apparent.

Zoe smiled, buoyed by their discoveries. "These change everything! With these files, we can..." She stopped at Paul's sharp intake of breath. He stood frozen, gazing at a single photo in the stack.

"What is it?" asked Zoe quietly. Paul did not reply but handed her the photo with a trembling hand. It showed Thompson flanked by two men, one of whom was an associate Paul recalled from his last unsolved case years ago. A name came unbidden to his lips in a haunted whisper. Gutierrez. How was Thompson connected to him? Just what deeper secrets did the victim's death now threaten to unearth from Paul's painful past?

When Paul found the photo of Gutierrez, Zoe noticed emotion had washed over him. There was a sense of familiarity and recognition as if a long-lost memory had been stirred. But there was also a surge of unease and apprehension. Knowing that Gutierrez's presence in the case could potentially unravel secrets from Paul's past, Zoe was unsure if she should say anything to Paul. It is a delicate balance, navigating the fine line between intrigue and trepidation. Zoe rationalized that as a retired detective, Paul knows that these unexpected discoveries often hold the key to unlocking the truth.

Despite the emotional turbulence, she knew, or hoped, he would remain determined to uncover the connection between Gutierrez and the mysteries that continue to haunt him. He felt Zoe's worried gaze but could not tear his own from the damning picture. Whatever answer it held, Paul knew this case had now become intensely personal. And its course would lead him directly to confront lingering ghosts that still haunted his dreams. With shadows from the past gripping Paul and new hope for breakthroughs stirring Zoe, they emerged more driven than ever to follow every twisting thread, no matter where the dangerous trail might take them next.

CHAPTER 3

SUSPECTS ABOUND

Paul stared down at the growing list of names and connections mapped out across the table. Each thread represented not just a potential killer but also tangled strands in the web that were Ray Thompson's life. Carlos Diaz, leader of a competing crime family, glared up from a mugshot. "Debts unpaid and disrespect shown," Paul mused.

Next to him, Simon Ames scowled from a driver's license photo. As a disgruntled investor that Thompson swindled out of millions of dollars, was revenge his motive? No love lost after a deal soured.

Further down the list, Alexa Chavez watched from her social media profiles. As Thompson's former girlfriend, what secrets did she hide behind perfect smiles? Frankie Morales, a local politician with possible ties to Thompson's company. Was he a crooked politician, or were those simply rumors spread by his opponents? Finally, Assistant District Attorney Tyler Greene peered out of some newspaper clippings chronicling his efforts to bring down Thompson and his illicit business dealings. What prompted Greene to suddenly back off his investigation? Did Thompson have something on Greene? With fortunes won and lost on ruthless deals, which battle was their last? Zoe studied the pieces with doubtful eyes. "It's like an octopus; cut off one arm, and three more pop up. How do we know where to start?"

Paul took a minute to respond. "We follow the trail of blood and money. One pathway will lead darker than the others. For now, all we have are possibilities. Our job is to make them probabilities, then certainties." His gaze lingered on Gutierrez's face—some certainties he dreaded to find. Let us go back to Thompson's home and talk to the housekeeper. Hopefully, she will be able to provide fresh information about the day Thompson was killed. Paul decided to drive this time. His car is a reliable old Ford sedan; nothing fancy or flashy, but it gets the job done. His philosophy is that it is not about the car; it is about the dedication and skills behind the wheel that matter. Additionally, Paul believes it is easier to blend in and go unnoticed when you do not draw too much attention to your vehicle.

When Paul offered to drive, Zoe looked at Paul's car and smiled. She thought it interesting that someone as respected as Paul chose to drive an old Ford sedan. It might not be a flashy choice, but it could say a lot about his character. Maybe Paul, she thought, values practicality over material possessions, or perhaps he is just attached to his trusty old car. Either way, it is those little quirks that make people unique.

After a 15-minute drive, mostly in silence, they pulled up outside the sprawling estate as Zoe peered out the window.

Admiring the estate, she said, "Money can't buy you love or a soul," she murmured. Zoe was curious about what the maid might know. In cases like this, Zoe has learned that the staff often have unique insights into the lives of the people they work for. She hoped that Martha, the maid, could shed some light on Ray Thompson's activities and the events leading up to his death. But she knew not to jump to conclusions just yet. It was important to approach the interview with an open mind and let the facts guide them.

Even after all his years of experience, Paul could not help but feel a mix of anticipation and caution. The grandness of the house made him wonder what secrets it might hold and what kind of person Ray Thompson truly was. He knew that interviewing the maid could be a crucial step in unraveling the truth behind his death. He also could not help but worry about Zoe's inexperience and how she would handle the interview. He hoped that she would stay focused and ask the right questions without being swayed by any potential distractions. As her mentor, Paul needed to guide her and ensure she stayed on track while still allowing her to learn and grow in her own investigative style.

Overall, he knew this interview was an important opportunity to gather information that could lead them closer to the truth. But there was also a sense of uncertainty, as he had come to expect the unexpected in the course of a complex investigation like this one. Martha answered the door with downcast eyes. "Please, come in. The master isn't..." Her voice caught, and she led them inside without finishing.

Martha, Thompson's maid, is a middle-aged woman, a St. Anne native, normally with a no-nonsense demeanor and a sharp eye for detail. She carried herself with an air of professionalism and efficiency that comes from years of experience in her line of work. Zoe was struck by her composed and observant nature. Her attire is always neat and tidy, reflecting her attention to detail. Martha seems to have a firm grasp on household affairs and possesses a strong sense of loyalty towards her late employer. One could tell she takes her job seriously and is not one to be easily fooled. There is an aura of mystery surrounding Martha, and Zoe is curious to unravel what role she might have played in this case.

Paul's impression of Martha was like Zoe's. She struck him as a reserved and diligent woman. Dressed in a crisp uniform, she had an air of professionalism about her. However, Paul noted her eyes held a weariness that spoke of years of challenging work. It was clear that she took her duties seriously and had a strong work ethic. Though she appeared slightly guarded, Paul sensed a level of loyalty within her.

Martha invited them into the kitchen, clutching a cup of tea, eyes flitting nervously. "I just cleaned. He liked everything clean, see."

Zoe smiled gently. "We know this is difficult, but we have to ask you a few questions. Let us start by you telling us about that night."

Martha swallowed. "He had guests. Loud voices, and then it went quiet. I did not think..." Her breath hitched. "In the morning, the police came. It is all my fault!"

Paul leaned forward. "We need the truth to find justice. Whatever you are afraid to say, we can protect you. Now, who were these guests?"

Fresh tears spilled from Martha as she crumpled. "Please, they will kill me too if they know I told!" Her desperation hinted at darker secrets within these walls. When Martha expressed fear and refused to disclose the names of Thompson's guests, Paul surmised it added a layer of complexity to the situation. It made him wonder what she might be afraid of and whether it was related to the case. Her fear could be indicative of her involvement or knowledge of something significant. It became clear that there might be more at stake than meets the eye, and it reinforced the notion that they were dealing with a tangled web of secrets and danger. Unraveling the truth would require delicacy and understanding, considering her fears and the possible repercussions.

Sympathetically, Paul said, "We will go for now; however, we will be back to finish our interview. Thank you for your time." Zoe was somewhat surprised Paul had ended the questioning so soon. They were just starting to ask questions to gain information from her about Ray Thompson on the night of his murder. But Zoe guessed Paul must have had his reasons for cutting it short. He has a lot of experience, far more than her, so maybe he sensed something off or saw a bigger picture that she could not see at the time. It was frustrating, but for now, she will trust Paul's instincts. She thought it better to not say anything to him.

Once outside, Paul suggested they visit Simon Ames next. Zoe simply nodded and got in the car. Her silence did not go unnoticed by Paul. She will get over it. It is all part of learning when an interview should be terminated. He made a mental note to talk to her about it later.

It was a short drive to Simon's home. But then again, living on a small island, everything is a short drive. Simon Ames' modest home is a quaint little place nestled in a quiet neighborhood. It is a cozy single-story house with a well-maintained exterior. The exterior is painted in a warm shade of beige, giving it a welcoming and unassuming appearance. Zoe was surprised, given the wealth that Simon had. Or maybe due to Thompson's swindle and Ames losing millions, this was as good as it would be for Simon. The front yard is neat and tidy, with a small garden filled with colorful flowers and a neatly trimmed lawn. The pathway leading to the front door is made of stone, adding a charming touch to the overall aesthetic.

They knocked on the front door and were greeted by Simon. No maid in this home. As they stepped inside, they found themselves in what could be described as a comfortable living room. The decor is simple yet tasteful, with a comfy-looking sofa, a couple of armchairs, and a wooden coffee table. The walls are adorned with a few pieces of artwork, giving the space a personal touch. The kitchen is modest in size, but it's equipped with all the essentials one would need. The countertops are clean and clutter-free, and there's a small dining table where Simon enjoys his meals. The kitchen gives off a warm and inviting atmosphere, making it a comfortable space to cook and gather. The bedroom it's a minimalistic haven. The bed is neatly made with crisp white sheets, and a small dresser sits against the wall. There's a window that allows natural light to filter in during the day, creating a bright and airy ambiance. Simon's home reflects his new, modest lifestyle. It's not extravagant or flashy, but it is a place that offers comfort and a sense of peace.

He invited them in; red eyes and crumpled clothes betrayed genuine grief. "We were not happy for a long time," Simon said quietly. "But I never wanted this. Ray did not deserve..." He broke off, choking back a sob. This was before Paul or Zoe even said hello.

Zoe asked gently, "What caused the troubles?" Paul smiled to himself, acknowledging Zoe taking the lead with the questioning. She may turn out to be a detective after all.

Simon sighed. "His work took over. Deals and clients, all hours. I felt neglected. But I would never harm him, even though he took most of my money. I am not that type of person. You must believe that!"

Paul studied Simon closely but saw only pain. Still, he pressed. "Where were you the night he died?"

Brow furrowing, Simon met his gaze steadily. "Here, alone." Paul nodded slowly, trusting the raw anguish in

Simon's voice and eyes. Loss unveiled the rawest truths, and this man's grief ran deep. For now, at least, his testimony rang true. More questions lingered, especially why Simon was so distraught over the death of someone who had stolen millions of dollars from him. Again, Paul cut the interview short, but this time, once outside, he asked Zoe what her thoughts were about Simon.

"I believe Simon's distress over Thompson's death was genuine. While it is true that Thompson swindled millions from him, emotions can be complex. People often have complicated relationships with those who have wronged them," Zoe speculated. Continuing, "In Simon's case, it is possible that despite the financial loss, there might have been some personal connection or history between him and Thompson that we are not fully aware of yet. People can hold conflicting emotions, such as anger and sadness, towards someone who has harmed them." Paul was impressed by her response, something that does not happen often.

Paul responded, "People can be quite skilled at putting on a show to deflect suspicion. However, genuine emotions can also be displayed in unexpected ways. In these situations, it is important to consider not only the emotional display but also the person's actions and motives. We should keep a close eye on Simon and gather more evidence before drawing any conclusions." Our next stop will be Thompson's office downtown. Downtown St. Anne is a vibrant and bustling area that serves as the beating heart of the Island. It is characterized by its charming blend of colonial architecture and Caribbean flair. As you stroll through the streets, you will find a mix of colorful buildings adorned with vibrant tropical flowers and local artwork. The narrow cobblestone streets are alive with the sounds of laughter and lively conversations. Sidewalk cafes and street vendors line the sidewalks, offering a variety of mouthwatering Caribbean cuisine, from jerk chicken to fresh seafood dishes. The savory aromas waft through the air, tempting your taste buds at every turn.

The main square of downtown St. Anne is a focal point, featuring a beautiful fountain surrounded by park benches where locals and tourists gather to relax and enjoy the warm Caribbean breeze. Nearby, you'll find a lively marketplace bustling with vendors selling all kinds of handmade crafts, artwork, and local produce. The downtown area also houses various small shops, boutiques, and galleries where you can browse unique handmade jewelry, clothing, and artwork. The vibrant colors and intricate designs reflect the rich cultural heritage of St. Anne. While the atmosphere is generally lively and inviting during the day, with the sun casting a warm glow on the town, it takes on a different charm in the evening. Soft lights illuminate the streets, casting a romantic ambiance. Live music can be heard spilling out of open windows and doorways, beckoning passersby to indulge in the rhythms of the Caribbean. Downtown St. Anne embraces a sense of community and is a hub of activity. Whether you are looking to explore the local culture, savor delicious cuisine, or simply soak up the vibrant atmosphere, this area offers a delightful experience that captures the spirit of the Caribbean.

Not exactly the setting one expects to find while investigating a violent murder. Zoe and Paul strode into the office building that housed Thompson's business. The building is an elegant structure that stands out amidst the backdrop of the town's quaint streets. It boasts an architectural style that echoes the colonial history of the

Caribbean with its grandeur and attention to detail.

Approaching the office building, you are greeted by a pristine facade adorned with ornate moldings and decorative accents. The exterior is painted in warm, inviting colors that harmonize with the surrounding tropical landscape. Tall windows with delicate ironwork frames allow natural light to filter into the interior, casting a soft glow on the polished wooden floors. The entrance is marked by a grand doorway framed by intricately carved columns, evoking a sense of sophistication and prosperity. Above the entrance, a sign proudly displays the name "Thompson & Associates," indicating the presence of a prestigious and established business.

Once inside, the atmosphere exudes professionalism and refinement. The reception area features tasteful furnishings, including plush chairs and a polished wooden desk, where a receptionist greets visitors with a warm smile. The walls are adorned with elegant artwork and framed accolades, showcasing the success and reputation of Thompson's firm. As you traverse through the building, you find well-appointed offices with spacious layouts, offering a comfortable yet professional environment for its employees.

Large windows provide picturesque views of the town, allowing ample natural light to fill the rooms. The decor is a harmonious blend of classic and contemporary design elements, reflecting Thompson's refined taste and attention to detail. The building also houses a conference room, complete with state-of-the-art technology for presentations and meetings. The room is adorned with a large table, comfortable seating, and tasteful decor, creating an ambiance conducive to productive discussions. Overall, Thompson's office building in downtown St. Anne emanates an air of prestige and success while still maintaining a sense of warmth and charm. It serves as a testament to the town's blend of traditional Caribbean heritage and the contemporary business world, creating a unique and inviting space within the quaint little town. Zoe thought it hard to believe that such a beautiful building could house such an evil person in Ray Thompson. They were met by directors Greg Adams and Ava Greene, who guided them to a plush office upstairs.

"A terrible tragedy," Greg sighed. "Ray was a mentor; his loss cuts deep."

Ava nodded, eyes dry. "If there's any way we can

help the investigation, please ask." Zoe once again proceeded with questions while Paul watched their faces. Beneath veneers of grief and cooperation, something seemed...off. A flash too quickly suppressed when certain topics arose. Zoe asked questions she thought were aimed at getting a better understanding of Thompson's background, relationships, and potential motives that could help in the investigation. "Can you provide any insight into Thompson's personal and professional life? Did Thompson have any known enemies or conflicts that we should be aware of?

Are there any suspicious or unusual circumstances surrounding Thompson's death? Have you noticed any recent changes in Thompson's behavior or activities? Are there any potential motives for someone to harm or want him dead? Ava answered first. "Mr. Thompson was a highly successful businessman with numerous business ventures and a reputation for being ruthless in his dealings. He had conflicts with other business partners and rivals, but I know of nothing specific that would directly link them to his death."

Ava also noted that Thompson had been acting more secretive in the days leading up to his death, which raised some suspicions. Greg Adams mentioned that Thompson had a reputation for being involved in shady business practices, but they didn't have any concrete evidence to support those claims. He also mentioned that Thompson had made many enemies over the years due to his cutthroat approach and questionable ethics. However, both directors denied having any information about the circumstances surrounding Thompson's death or any firsthand knowledge of his enemies. Zoe knew that, as a detective, she needed to approach all information with skepticism until it was verified. While she didn't fully believe their answers without further investigation, their insights provided valuable leads that needed to be explored. It was important to corroborate their information with other sources and evidence to validate or invalidate their claims.

As they stood to leave, Paul said casually, "Just one thing. The night he died, where were negotiations at with the Ortiz contract?" Greg froze, exchanging a loaded look with Ava.

"I, uh, apologies, detective. That deal was actually moved to next week if I recall correctly."

Paul smiled thinly. "Funny because our records state

it was scheduled for that night. An easy thing to mix up, of course." Their masks cracked for an instant, revealing something harsher in the depths. No, Paul thought, these two knew Thompson far better than they let on. More digging was required here amongst the glass and gleam. Once outside, Zoe had to ask, "What is the Ortiz contract, and why is this the first time I am hearing about it?"

Paul knew it was coming, "It refers to a confidential agreement between Ray Thompson with a certain Mr. Ortiz. The details are still murky, but it seems to involve some shady business dealings and potentially illegal activities."

Zoe thought before speaking, "It caught me off guard, not knowing anything about it."

Paul's response was typical for Paul, "Sometimes information comes up that we did not anticipate." Zoe chose wisely to drop the conversation. She decided that Paul had a lot of experience under his belt, and he knew how to handle these kinds of cases. She needs to trust him to make the right calls. Besides, she is still learning the ropes, so she viewed it as a learning opportunity. They got in Paul's car and drove back to the police station. Upon arrival, the desk sergeant told Paul and Zoe that they had arrested Maxwell Grant on an old warrant. Grant was handcuffed and waiting in an interrogation room, asking to see Paul. Maxwell Grant was one of Thompson's largest investors and fiercest rivals. Zoe had heard the name but never had any dealings with him.

Maxwell smiled slowly at their entrance. "My, if it is not the intrepid detectives. Come to offer me a deal, I presume? I have information, you see. Things that could... complicate matters for certain associates of the deceased."

Paul was very familiar with Maxwell Grant. Grant was a complex and enigmatic individual. He possessed a certain air of mystery that made it challenging to fully understand his true motivations and intentions. He exuded quiet confidence, always remaining one step ahead of the game. He had a clever mind, using his intelligence and cunning to manipulate those around him. In terms of appearance, Maxwell Grant cuts an intriguing figure. He is of average height, with a lean build that suggests a disciplined lifestyle. His sharp features seem chiseled, giving him an almost predatory look. His piercing eyes, often hidden behind a calculating gaze, reveal a keen intellect and a shrewd personality.

Grant moved with a deliberate grace, his every action

measured and thought out. He had a way of commanding attention without raising his voice or drawing unnecessary attention to himself. There is an air of danger around him, an underlying sense that he is capable of great harm. Despite his composed exterior, there is something unsettling about Maxwell Grant. He carried an aura of darkness as if he had seen and done things that most would find unfathomable. There is an aura that surrounds him, hinting at a troubled past and a complex web of secrets.

Zoe shifted uncomfortably under his oily stare. "What are you implying?"

Maxwell leaned forward. "Let's just say Thompson had a lot of secrets. Dark things, the kind that gets good men like yourselves into trouble should they be exposed. Unless, of course, I was to receive full immunity for my cooperation..."

Paul scrutinized the man coldly. "Blackmail will get you nowhere, Mr. Grant. If you have legitimate leads, share them freely. Otherwise, I suspect your information is worth less than your soiled word."

For a moment, fury and something more sinister

flashed in Maxwell's eyes. "You will regret ignoring me, detective. Mark my words..." Paul and Zoe left the interrogation. Realizing it was getting late and still needing time to digest what Maxwell had said, they decided to order in and review the day's events. Over cartons of lo mein, Paul filled Zoe in on what he knew about Maxwell and the mysterious Ortiz contract.

"Maxwell was inches from losing the Ortiz deal to Thompson. However, the night of the murder, Ortiz and Maxwell finalized the deal instead. However, Maxwell's alibi holds, so we can't rush to conclusions. Not yet."

Zoe was stunned. How did Paul come across this information? Why had he not shared it with her? This time, she had to say something. "I believe in open communication and collaboration. I do not see the need to intentionally withhold information from me. I understand the importance of sharing information to solve the case effectively. It is frustrating when I find out important information during interviews with suspects instead of you sharing it with me beforehand. It makes me feel a bit left out and like I'm playing catch-up."

Paul was waiting for this. "I don't intentionally

withhold information from you. It is more a matter of strategy. Revealing certain details to a suspect during an interview can elicit genuine reactions, helping us gauge their honesty or possibly catching them off guard. It is all about maintaining the element of surprise and playing our cards wisely. In an investigation such as this, timing and control are crucial. I only have our best interests at heart when I choose when and what to share." Paul continued, "I only came across this information recently, and we are moving in so many different directions I do not have time to share everything I came across."

Zoe's face fell. "I'm trying to understand, but it is not enough, is it? I'll never measure up to your experience."

"Don't be too hard on yourself. Investigation is as much an art as science," Paul reassured. "What is important is openness to doubt and new facts. Your initiative shows promise; just slow down and scrutinize all sides. 'Why' is every bit as key as 'who.' "She nodded; fire rekindled in her eyes. "Then help me understand so next time I have a tougher skin." Her eagerness, while premature, gave Paul hope this case could challenge them both for the better. Paul suggested Zoe continue researching Thompson, Grant, and the other suspects they had identified.

Paul called it a day and left Zoe to do her research. It was getting to the point that he could no longer manage these long days. And nights. He got in his Ford and drove home, eager for his hammock and ocean view. After a long day of detective work, Paul usually finds solace in the simple pleasures. With a glass of his favorite rum with coconut, he sat on his veranda and watched the sun sink beneath the horizon. The rhythmic sound of waves lapping against the shore brings a sense of calm. At times, he will delve into a good novel, immersing himself in the complexities of another world, if only for a little while. And lately, he has been contemplating boatbuilding. He loves working with his hands, which he believes will allow him to find peace amidst the chaos of the day. Meanwhile, Zoe emerged from her research with troubling news. Thompson owned shell companies in Panama and the Caymans. Digging deeper uncovered suspicious transactions payoffs disguised as private equity deals. She had to call Paul.

When the telephone rang, and Paul saw it was Zoe, he flinched. When he is trying to find a respite from the day's events and indulge in some well-deserved relaxation, a call from Zoe regarding work-related issues can be quite...disruptive. It inevitably stirs up a mixture of emotions. On one hand, there is a sense of duty and responsibility to assist in any way he can. After all, he had taken on the role of mentor to Zoe, guiding her in this treacherous world of investigations. On the other hand, there is a tinge of frustration, a longing for uninterrupted solitude. It is a constant battle between his desire for peace and his commitment to justice.

The mysteries of the world have a way of creeping into even the most idyllic moments, reminding him that the shadows are always lurking nearby. Paul answered the phone a bit annoyed but considered the information Zoe relayed to him thoughtfully. Corruption is a strong motivator, to be sure. Powerful enemies could lurk in Thompson's shadow deals. But why kill him now after so long untouched?

She leaned forward eagerly, continuing, "Someone may have been blackmailing him with this evidence, or he threatened to expose an accomplice and had to be silenced. It gives us real motive beyond petty disputes." Paul had to agree. His mind was turning.

"You may be onto something, Zoe. If done right, illegal empires can topple as easily from within as without. I know people who can unearth more in these offshore situations. Good work, keep following this thread, and others may unwind." Their first solid lead was beginning to cast the case in a darker light indeed.

The next morning, Paul drove to the station to find Zoe already there. "Have you been here all night?"

Zoe looked a bit worse for wear. Her eyes were bloodshot, and she had some serious under-eye bags going on. She had a disheveled appearance, with her hair all over the place. Paul was smiling at her, trying to conceal his amusement at her appearance. His response to the way she looked put a smile on Zoe's face. She thought it nice to have a moment of levity amidst all the seriousness of the case. Keeps things in perspective. Paul brewed strong coffee as Zoe flipped through her notes, frustration etched across her brow. "It is like they are all lying," she muttered. "How do we find the truth?"

Paul handed her a steaming mug and gazed out the window, considering. "Doubt is our strongest ally in this game of deception. While suspicion clings to many, trust none fully. Seek the lies within truth and truth within lies. Don't judge by assumed motives but by the facts alone. Our duty is to the victim, not vengeance." He turned to her with a soft smile. "Rome was not built in a day, and neither are cases won. Have faith that answers exist and patience that they will come clear in time. A new day may shine new light through these darkened paths we tread."

With that, Paul gathered Zoe's coat and told her to go home, get some sleep, and return the next day, leaving her to ponder his wise words as she reviewed their findings once more under the window's slowly lightening sky. Somewhere, the pieces of this twisted puzzle were falling into place, and she vowed to be ready when they did.

With their first day of questioning behind them, Paul and Zoe felt no closer to answers in the vexing case. Thompson's life seemed tangled in deception as much as the threads of their inquiry. Yet, through patient work and partnership, kernels of truth had been unearthed from the murky details of his past. As they departed for the day, both detectives felt their understanding had grown, with hopes that tomorrow's revelations would bring further light. Though the path ahead remained dim, their shared resolve to do justice lit the way forward into the mystery's deepening shadows.

CHAPTER 4

UNTANGLING THE WEB

After leaving Zoe and hoping she would go home to get some sleep, Paul went home with the intention of doing nothing. Instead, Paul pored over bank statements and transaction records late into the night. Something was not adding up with Thompson's finances. He noticed a pattern of large cash deposits followed by wires to offshore shell companies. Investigating offshore accounts and wire transfers is no easy task; Paul learned that years ago. However, it is much easier now than before the advent of the Internet.

Paul follows the same process each time: The first step is to gather as much information as possible about the individuals or entities involved. Look for any suspicious financial activities, connections to known tax havens, or complex transactions that might raise eyebrows. Next, dive into the paper trail: Request bank statements, transaction logs, and any other relevant financial documents. Follow the money, so to speak, tracking each wire transfer and offshore account to uncover any hidden connections. It is often a game of cat and mouse. Those involved in these activities are aware of the scrutiny they may face and take precautions to conceal their tracks. That is when attention to detail and investigative skills come into play. Cross-reference the obtained records with other sources such as company filings, legal documents, and even international financial databases.

Look for inconsistencies, hidden beneficiaries, or unusual movement of funds. Paul learned long ago it is all about connecting the dots and following the threads of evidence. Paul also learned that it is not a straightforward process. There may be legal hurdles, jurisdictional challenges, and privacy concerns. But with persistence and sometimes a little help from international law enforcement agencies, you can unravel the web of offshore accounts and wire transfers and expose those trying to hide their ill-gotten gains. It is not as glamorous as it seems, but the satisfaction of bringing the truth to light is worth it. In fact, for Paul, it is the least liked part of his job. Boring, in fact.

The following day, Paul called Zoe at home and asked what time she would be in the office. "I will be there within the hour," Zoe replied, hoping she would make the self-imposed deadline.

Zoe looked around her apartment to ensure she had everything she needed to take with her. Her apartment can be described as 'cozy.' When she was apartment shopping, she insisted it not be too far from the water. She can catch glimpses of the sparkling Caribbean Sea from her balcony. Like most residents of St. Anne, she loves the calming sound of the waves crashing against the shore. It is the perfect location for her, close enough to the action but also providing that peaceful retreat when needed. She was very firm about the type of apartment she wanted, and this apartment fit her every need. The living room is both comfortable and functional. The space is tastefully decorated with shades of blue and white, giving it a coastal vibe. There is a plush couch where she can kick back and relax after a long day of detective work. The walls are adorned with posters of some of her favorite mystery novels and framed photographs of friends and family. Moving towards the kitchen, one will notice sleek countertops and modern appliances.

She loves to cook, so her kitchen is well-equipped

with everything needed to whip up a delicious meal. Her favorite part of the kitchen is a small dining nook where she can enjoy morning coffee or share meals with friends when they come over. The bedroom is her sanctuary. It is adorned with soft linens and has a calming color palette. The natural sunlight streaming through the window adds warmth to the room. She has a desk in the corner, complete with her trusty laptop, where she can analyze case files and dig deep into investigations. Overall, Zoe's apartment is a reflection of her personality, practical and modern.

When she arrived at the police station, she found Paul sitting at her desk, waiting. She apologized for being late, although she was within the hour timeframe she set. She sat in a chair, allowing Paul to remain at her desk.

Paul shared his findings with Zoe over coffee. "Look at these money trails. Thompson receives monthly cash deposits of fifty to a hundred thousand dollars. Then a few days later, the same amount gets wired to companies in the Caymans or British Virgin Islands."

Zoe frowned. "That sounds suspicious,"

"But these amounts are too small for Carlos Diaz's

scale of operations. My guess is someone was paying Thompson for influence and access. Maybe a competing business or dirty politician."

"So, blackmail then. Do you think whoever it was silenced Thompson before he could spill their secrets?"

Paul took a thoughtful sip of coffee. "It's looking that way. Our victim had his hands in a lot of pots. With that kind of leveraged information, he must have ruffled some very powerful feathers."

"So now we just need to follow the money and see where it leads." Zoe leaned in eagerly.

"This could be our big break." Paul hatched a plan to trace the money offshore. I know a guy at St. Anne Bank who handles a lot of shell company transactions. If we go undercover as investors, maybe we can get him to reveal more. The thought of undercover work excited Zoe. She thought undercover work can be exhilarating, immersing oneself in a different identity and infiltrating a world of secrets and deception. It offers a unique perspective and the opportunity to gather crucial information that might be inaccessible otherwise. This assignment, however, should be a piece of cake, she thought. Meeting with a bank executive did not seem all that exciting.

The next day, dressed in expensive suits, Paul and Zoe drove to St. Anne Bank with Zoe at the wheel. Her sleek black sedan is dependable and inconspicuous but certainly a better option than Paul's old Ford. They walked into the bank and dropped a few names to ensure they had met with the senior banker they needed to speak to. David Chen promptly greeted them.

David Chen is a tall, distinguished man, always impeccably dressed in tailored suits that scream success. He has a commanding presence and an air of confidence that comes with his years of experience in the banking industry. His dark, slicked-back hair and sharp features give him an almost intimidating aura. As a senior banker, David is known for his shrewd business acumen and astute decision-making. He is a master at navigating the complex world of finance, always one step ahead of his competitors. Many admire his ability to analyze risks and seize opportunities with precision and finesse. While Chen might come across as reserved and professional, beneath that polished exterior lies a highly driven individual. He is fiercely ambitious and constantly seeking new ways to expand his influence and wealth. Some might say he is ruthless in his pursuit of success, willing to do whatever it takes to achieve his goals.

In the banking world, David Chen is a formidable force to be reckoned with. His connections, knowledge, and wealth have made him a respected figure, both feared and admired by his peers. He is a man who knows the price of power and is willing to pay it. But like any human, Chen has his flaws, too. Behind closed doors, there is a vulnerability that he rarely shows to the world. He is haunted by his own demons and carries the weight of his past choices. The corridors of high finance can be a treacherous place, and David has had to make sacrifices that still weigh heavily on his conscience. Over coffee in his office, Chen laid out services for private clients.

Paul gestured casually at their wristwatches. "We are diversifying some assets and want more... off-the-books opportunities. Know what I mean?"

A guarded look entered Chen's eyes. "I understand certain clients value discretion. Let me pull up some options in the Caymans or British Virgins." As he typed, Zoe noticed bank records on his computer. She mentioned an interest in

Thompson's portfolio. Chen froze. "I... I am not actually involved in Mr. Thompson's accounts." A sheen of sweat broke on his brow as he realized his mistake.

Before he could backtrack, Paul pressed him. "We just want the same advantages our associates enjoyed. No need to cause trouble, right David?"

The banker swallowed anxiously. "If you will excuse me, I just remembered an urgent call. Please, let me show you some alternatives another time." With that, David Chen got up and walked out of his office, leaving Paul and Zoe sitting by themselves.

Paul looked at Zoe, smiling, "I guess we should leave as well?"

Back at the station, Zoe searched public records for the shell companies. She gasped as connections began forming on her screen. "Paul, come take a look at this."

He peered over her shoulder, eyes narrowing as familiar names jumped out at him. "Those holding companies, they came up in a case I worked years ago, looking into political graft. Same money laundering setup, only the names had changed." Zoe swiveled her chair to face him. "What are the chances this is all somehow connected to your old case?"

Paul ran a hand over his face, memories of that painful time swirling in his mind. "Too many to be a coincidence. That case haunted me; I was so close to busting the whole corruption ring when the leads suddenly dried up. My contacts vanished overnight."

"So, do you think Thompson's murder was somehow related to covering up whatever you were initially investigating?" Zoe prodded gently.

His dark expression gave no answers, but the wrathful gleam in his eyes spoke volumes. "Whoever did this will regret resurrecting old ghosts. It's time I finished what I started. The truth will turn out, one way or another."

So, with a new focus from the financial leads, Paul and Zoe set out to re-examine Thompson's circle. They started with Maxwell Grant, given his history with the victim.

Grant and Ray Thompson were more than just acquaintances. They had a long-standing business partnership spanning years, tied together by shady dealings and shared secrets. Grant was like a shadow, always lurking in the background, quietly managing the intricacies of Thompson's business empire. Thompson relied heavily on Grant's expertise in maneuvering through the dark underbelly of the business world. Whether it was concealing financial discrepancies or orchestrating questionable deals, Grant's role was pivotal in maintaining Thompson's empire. Together, they amassed wealth and power, untangling webs of deceit in pursuit of their shared ambitions.

The drive to Grant's estate, located on the far side of the Island, away from downtown, was along one of the most scenic and beautiful roads on St. Anne. The estate is a stunning retreat that embodies both luxury and tranquility. Located on a secluded beachfront, Grant's home was nestled amidst lush tropical foliage and swaying palm trees. The property exuded a sense of privacy and exclusivity, offering a serene escape from the outside world—fitting for a crook like Grant. The architectural design is a harmonious blend of contemporary elegance and Caribbean charm. The exterior features crisp white walls accented with vibrant pops of color from the local flora and fauna. Expansive windows and open verandas allow for breathtaking panoramic views of the crystal-clear turquoise waters.

Stepping inside, one is greeted by a light and airy ambiance.

The interior decor is tastefully minimalist, capturing the essence of island living. Polished hardwood floors, coupled with whitewashed walls and high ceilings, create a sense of space and serenity. The living area is adorned with plush, comfortable furnishings, inviting relaxation and conversation. Large floor-to-ceiling windows bring the outside in, offering glimpses of the shimmering ocean and golden sandy beaches. Grant's home boasts a well-equipped gourmet kitchen, perfect for preparing culinary delights using locally sourced ingredients. A spacious dining area with a large, rustic wooden table is set against panoramic views, providing an unforgettable backdrop for intimate meals.

The bedrooms exude a sense of luxury and comfort. Crisp linens, plush mattresses, and soft lighting create a tranquil haven for rest and rejuvenation. Each bedroom showcases its own private balcony or terrace, offering breathtaking views of the surrounding landscape.

Outside, a beautifully landscaped garden surrounds the property, with a variety of tropical plants and colorful flowers. A pristine infinity pool stretches out towards the horizon, providing a serene oasis for swimming and lounging. Poolside cabanas and a sun-kissed terrace offer the perfect setting for soaking up the Caribbean sun or enjoying cool evening breezes. Upon seeing Paul at his door, Maxwell, out on bond, dropped his smug veneer.

"Detective, this is unexpected. How can I help you?"

Paul smiled coldly. "We appreciate your cooperation, Mr. Grant. Can we come in to discuss your dealings with Thompson?"

Sweat broke on Maxwell's brow, but he gestured them inside. Once the door closed, Paul rushed Grant, trapping him against the wall. "Let's skip the games. I know you two were in cahoots, so start talking before things get unpleasant."

Grant trembled under the steel in Paul's eyes. "Alright, alright! Ray was blackmailing me; he said he would tell my wife about our affairs unless I played ball." Details began pouring out of Grant's mouth faster than Zoe could make notes about skewed bids and kickbacks.

Paul released him in disgust. "Anything else?"

Grant told them all he knew, terrified of incurring the wrath of this legendary detective from his past. During the intense exchange, Maxwell Grant revealed some key details about their association. He shared information about the intricate financial schemes they were involved in, the extent of their collaboration in illicit activities, and the specific ways in which Thompson's empire was intertwined with Grant's own endeavors. He also disclosed the motivations behind their partnership, shedding light on the shared secrets and circumstances that bound them together. He revealed the lengths they went to protect their interests and maintain their power, potentially divulging hidden connections to other high-profile individuals and organizations.

Grant also disclosed the specific actions they took to further their goals, exploitations, and manipulations they engaged in, as well as the steps they took to cover their tracks. These revelations would likely shed light on the complex web of deceit and the true nature of their association. Paul sarcastically said thank you, and then he and Zoe left Maxwell Grant, still standing against the wall.

Armed with fresh intel, Paul and Zoe hit other leads, leveraging his fearsome reputation to squeeze witnesses like oranges. By day's end, they had pieced together Thompson's intricate web of corruption. Paul had enough of being Mr. Nice Guy with the suspects. He was retired, so his job would not be in jeopardy if someone were to complain about his tactics. He impressed upon Zoe and said that she should in no way follow his lead.

With leads drying up through legal means, Paul decided drastic actions were needed. They needed to make another trip to Thompson's office. That night, he and Zoe stealthily broke into Thompson's office building. This time, Paul drove so Zoe's car would not be seen at the break-in. Paul picked the lock on the front door, and once inside, they went straight to Thompson's office. They started searching desks in the dark, flashlight beams dancing around documents. Zoe's light froze on tapes of security footage.

Playing the tape, Zoe excitedly said, "Paul, look!" On the screen, a masked figure slipped into the office, according to the date stamp, days ago. They watched intently, blood running cold as the intruder rifled through files, unaware he was being recorded.

A creak outside snapped their focus. Voices grew louder as guards approached for rounds. Paul dragged Zoe behind a cabinet as boots echoed past. They held their breath until the voices faded. Once clear, they resumed searching feverishly. Paul broke the lock on Thompson's personal safe, yanking out bundles of cash and a small notebook. Flipping pages under his light, Zoe gasped. Entries detailing transactions and blackmail material, their killer's motive was spelled out within. They had struck gold, but daylight was approaching. Paul replaced the book and relocked the safe. It was time to vanish before the next guard walked by.

Back at the station, evidence from the night's exploits stretched across whiteboards. Financial records mapped a web of corruption with tendrils in multiple countries. Zoe leaned wearily against a desk, rubbing her eyes.

"How are we ever going to untangle this mess? We are up against mob bosses, politicians, cartels... I am out of my depth here."

Paul patted her shoulder bracingly. "I know it is a lot. But we have already come so far, piecing things together. Have faith in the process."

Doubt, however, continued to gnaw at her. "What if we miss something crucial? Make a mistake that costs lives? I feel like a rookie, but this case is much bigger than me."

Gripping her arm, Paul forced her gaze up to meet his steady one. "Listen to me. You have a gift for this job; do not sell yourself short. I have seen your drive, your instincts. Have heart, you were born for challenges like this."

Zoe took a steadying breath, squaring her shoulders back. "You're right. Giving up will not help Thompson or bring his killer to justice. I am staying in this fight until the end. Now, what is our next move, boss?"

Paul smiled when Zoe said "boss," with renewed vigor in both their eyes. "Let's pay a visit to our mysterious intruder.

In the late hours, Paul and Zoe were burning the midnight oil once more. Zoe rubbed her tired eyes, lamenting how much ground they still had to cover. Catching her exhaustion, Paul suggested a break. "Come on, first rounds are on me. I know a great dive near here."

The 'great' dive is Paul's hidden gem tucked away in the corner of the island. The place had a certain charm to it, dimly lit, with weathered wooden stools and tables. The walls were adorned with nautical memorabilia, and the air carried the scent of salt and aged rum. It is the kind of place where locals gather, seeking solace in the comfort of their drinks. Not exactly glamorous, but there is an authenticity to it that Paul appreciates—just the kind of spot to unwind and let loose after a long day of unraveling mysteries.

The tiny bar was empty, save a few regulars. They claimed a corner booth, nursing their drinks in the low light. Paul ordered a draught beer, and Zoe ordered a classic mojito. She enjoys the refreshing combination of mint, lime, and rum, as it always hits the spot. After the drinks arrived, Paul sighed. "You know, it was not always easy for me either when I started out. Solving my first big case nearly got me killed."

Intrigued, Zoe leaned in as he launched into the harrowing tale. Her eyes widened at his close calls, narrow escapes. She never knew the legends had feet of clay, too, once. Paul smiled wistfully at the memories. "You learn from your failures. Get back up and try a new angle. Before I knew it, I was the one providing backup to the rookies." Renewed vigor filled Zoe as she grasped his hand gratefully.

"Thank you for believing in me like your colleagues must have believed in you. I promise I will keep pushing through til we nail this son of a bitch. For Thompson and us."

Glancing at their joined hands, Paul nodded back with pride in his eyes. "For justice. Now, let's get some rest. We have work to do tomorrow."

Refreshed after a night's rest, they met for breakfast at Zoe's favorite breakfast spot, a cute little café called "Sunrise Bites." They serve the most delicious tropical fruit platters with a side of freshly baked pastries. The aroma of coffee permeates the air as you enjoy your breakfast overlooking the stunning beach. They dove back into theorizing over breakfast. Trading hypotheses fueled more coffee refills as the morning slipped away.

Zoe leaned back with a thoughtful rub of her chin. "So, in your opinion, Maxwell Grant is just a greedy pawn, and we should focus our attention higher up the food chain, yes?"

"I'm inclined to agree," Paul mused. "But do not think flattery will distract me from the real suspects, no matter how charmingly you employ it, Miss Walker."

She laughed, cheeks pinking. "Keep your eyes on the prize, old man, lest you miss what is right under your nose. Like how Ames' alibi seems paper thin upon further scrutiny."

"Careful now, do not let that vivid imagination run away with you," he chided playfully. "For all your sass, you have the makings of a fine detective. It has been an honor working alongside you."

Zoe smiled softly. "The honor is mine. Though between your tales and distraction tactics, I fear our culprit will walk free by week's end at this rate."

"Better wrap this up fast then," Paul winked. "I would not want your talents to be wasted on lesser cases once I am gone." With refreshed cheer and camaraderie, they dove back into theorizing with renewed vigor. As evening fell, Paul and Zoe finished organizing leads for the next day. Zoe yawned but smiled brightly, filled with renewed energy from their progress.

"I think we are really close to a breakthrough. A few more interviews and something is bound to crack."

Paul returned her smile, but it did not reach his eyes.

His mind remained heavy with ghosts of the past unearthed by this case: so many secrets yet to be uncovered, so many old wounds still raw beneath the surface.

Noticing his dark mood, Zoe squeezed his arm gently. "Do not lose heart. Whatever demons from your past are crawling back, we will face them together. I have got your back."

He nodded, grateful for her support in this stormy present. "Get some rest. Tomorrow, we end this, one way or another."

With hope guiding one detective and shadowing the other, their fates remained intertwined. The end was closing in, but what darker truths would be exposed along the way?

CHAPTER 5
PHANTOMS OF THE PAST

St. Anne in the morning, the mere mention of morning in the Caribbean evokes tranquility.

As the first light peeks over the horizon, a gentle warmth washes over the island. The cerulean sky begins to come alive with hues of pink, orange, and gold. The air is still and crisp, carrying the faint scent of saltwater and tropical blooms. The sound of waves kissing the shore lulls you into a peaceful state of mind. It's a moment of serene beauty, a sight that reminds you why this place is called paradise—simply magical.

Working on a tough case can be quite the contrast to the peaceful mornings in the Caribbean. While the sunrise serves as a reminder of serenity and tranquility, a difficult case brings forth a storm of emotions and challenges. Gone is the calming stillness, replaced by a sense of urgency and determination. The complexities of the investigation and the weight of uncovering the truth can disturb the tranquility of the island. There are moments of frustration as Paul finds himself delving into the dark and murky depths of the human psyche, all while surrounded by the beauty of nature. It becomes a constant battle, balancing the quest for justice with the longing for peace.

Yet, amidst the chaos, there is a glimmer of hope that Paul can restore order and make a difference. The contrast is stark, but it is the very challenge that keeps him grounded and reminds him of the importance of his work. Another sunrise ruined.

Paul sifted through the stacks of files on the desk, piecing together Ray Thompson's shadowy financial dealings. Something on the page caught his eye: a faded driver's license photo. His blood ran cold as memories came flooding back. The face staring out was Gutierrez, a low-level smuggler the police had been tracking for at least ten years. Gutierrez had given them the slip, disappearing amid rumors of a bigger player pulling strings. The case has haunted Paul ever since. Unanswered questions about whether more lives could have been saved if they had caught the true mastermind still plagued his dreams. Now, here was Gutierrez's face, connected once more to corruption.

Paul rubbed his temple, feeling the beginnings of a migraine. He did not want to relive this ghost from his past. But his past was not done with him yet. Gutierrez was the missing link between this new mystery and the one that had driven Paul away from the job years ago. Gutierrez, a name that sends shivers down Paul's spine. He has proved to be a formidable adversary, a figure from Paul's past that haunts his every step. Gutierrez is a cunning and manipulative individual, always one step ahead of the game. He displays an aura of darkness, leaving a trail of chaos and destruction in their wake. This person possesses an unsettling ability to blend into the shadows, making it difficult to discern his true intentions. Whether it be his calculating mind or skillful deception, Gutierrez is a master of manipulation. He knows how to exploit weaknesses, how to exploit the fragility of a person's past and use it to his advantage.

Gutierrez revels in chaos, thriving on the power he gains from it. He leaves no stone unturned in their pursuit of their own twisted agenda. His actions are driven by a twisted sense of justice, often blurred by his own personal vendettas. Paul's encounters with Gutierrez have always been a battle of wits, a dangerous dance on the thin line between life and death. They are a constant reminder of Paul's past failures and shortcomings. Paul had been determined to bring Gutierrez to justice and put an end to his reign of darkness once and for all. If Paul wanted answers for Thompson's death, he must confront the demons he thought were long buried. Paul steeled himself for the digging he knew he must do. The truth was out there, and this time, he would be ready for whatever horrors it exposed.

Paul closed his eyes and leaned back as memories washed over him...He was undercover at a dingy warehouse, tracking a shipment set to arrive. A man matching Gutierrez's description oversaw crates being loaded from a truck. Paul grew tense; something did not feel right. As night fell, his radio crackled, and they lost the truck. Swearing under his breath, Paul followed Gutierrez into an office, gun drawn. But he wasn't quick enough. A sack was thrown over his head as bodies crashed into him. He struggled vainly as duct tape wound around his wrists. A shadow loomed in the darkness.

"You thought you were clever, detective. But I'm always one step ahead." Gutierrez's icy laugh sent chills through Paul. They dumped him bound and blind on the dock hours later. By the time backup cut him free, the entire operation had vanished without a trace. Paul was left with nothing but ghosts of might-have-been. If only he had been faster, smarter. Lives could have been saved from the devastating cargo that had surely reached its destination. The failure gnawed at him, driving him from the force.

Now Gutierrez's specter had returned to dredge it all up again. Paul opened his eyes with a jolt. The past receded as he returned to the present. Zoe watched him with concern.

"Are you alright? You went somewhere dark for a moment." Zoe has mixed feelings about Paul and his 'dark place.' She understands Paul's past trauma can influence his work and lead him to a dark place. It is tough for anyone to carry the weight of their past, especially when it involves some dark mysteries. But at the same time, it can be frustrating because it affects their investigation. She needs Paul to be focused and clear-headed to crack this case. Zoe tries to be patient and supportive, offering him the guidance he needs to navigate through his past without completely losing sight of the present. It is a challenge, but Zoe believes in Paul and his ability to overcome his demons.

He rubbed his eyes, banishing the lingering shadows.

"It's nothing. Old memories, that's all."

But Zoe was not so easily dismissed. "You have been off ever since finding that photo. What are you not telling me?"

Paul knew she would not let it go. Still, he was not ready to relive that failure aloud. "It is personal. Let us just say this case opens some old wounds." Zoe frowned but did not press further.

As they worked, Paul found himself drifting away again. His mind kept replaying that night, looking for what he had missed. If only he had seen the signs and been faster. How many lives had been lost or ruined because of his failure? And now, his ghosts had not finished haunting him yet.

Zoe shot him concerned looks, but he ignored her worries. To solve this case, he would have to confront his demons head-on, even if they threatened to drown him in the past once more.

Paul dove into records of Gutierrez's past appearances. A connection emerged that made his blood run cold. The last sighting of Gutierrez was in a border town, the same place where Paul's botched operation was meant to intercept a shipment ten years ago. Drilling deeper, Paul uncovered more ominous threads. Witnesses reported Gutierrez frequenting a warehouse near the docks. Continuing the tedious work of examining financial records showed payments from the same shell companies Paul had been tracking a decade back, now shuffling money through Thompson's empire.

It was all connected, this man, Thompson, even Paul's long-buried failure. Gutierrez and whoever pulled his strings had evaded justice for ten years. And now Thompson's death threatened to expose whatever secret world of corruption they'd been spinning webs in. A cold fury gripped Paul. Too many horrific possibilities swirled in his mind. How much suffering had been deliberately inflicted just to cover greedy tracks? This time, there would be no slipping through the cracks. Paul was closing in, and he would be damned if any demons escaped him again. Paul sank into a chair, thoughts swirling. So much matched, it couldn't be a coincidence.

But did he dare believe this could finally answer the ghosts haunting him? Dare hope he might find the resolution stolen from him a decade past. Or would the past only drag him under again? Old wounds were reopening, and he did not know if he had the strength left to face what lay below the surface.

Zoe watched him wrestle his demons, concern clear. But she knew words were useless for the private battle within. All she could offer was silent support until he found his way clear again. Paul stared unseeingly at the files, re-living each piece and clicking into the jigsaw of his failure. Anger and doubt warred within. If he pursued this, would the truth set him free at last? Or open new cages from which there was no escape? Only by confronting the darkness could he ever truly leave it behind. He had come this far; he had to see it through to the end. Paul delved into dusty boxes in the evidence locker. His gut told him the answers lay in the past.

Concentrating on decaying financials, one entry stopped him short—a huge payment to Gutierrez two weeks before Paul's disastrous bust. From the same offshore account, it was later tied to Thompson. Heart racing, he pulled their target lists. Sure enough, Gutierrez's circles overlapped with the figures Paul had been chasing a decade ago. The spider's web was larger than they had realized. A shipping manifest revealed the size and destination of the "lost" cargo from Paul's operation. It matched another document plugging leaks in Thompson's empire.

Dates and ports of call aligned like constellations in the night sky. Individual stars took form as part of a hidden picture. Piece by piece, the cases merged in Paul's mind. After ten years, the elusive puppet master had finally come into focus. All paths led back to one man weaving the threads from shadows. Paul's failure and Thompson's death were moves in his long game. Breaking the surface at last, stale air filled his lungs. He had found what he sought, but now he just had to make the puppets' puller reveal himself. Paul emerged from the evidence locker deep in thought.

Zoe had waited, sensing his need for solitude in the hunt. "Find what you were looking for?" she asked gently.

Paul nodded. "The pieces are falling into place." He told her of the connections emerging between cases.

Zoe listened intently, admiring his growing tenacity. "I am glad you did not give up. Closure is so important." Her somber tone hinted at a deeper meaning.

"I need some fresh air. And a drink," Paul said.

"Want to join me?"

Zoe drove back to Paul's 'dive' bar. Over drinks, Paul probed gently, "Why do you do this job?"

Zoe sighed. She spoke of a childhood crime that rocked her small town, the victim a close friend. Questions lingered for years with no answers. The crime was the disappearance of a young girl named Sarah. She went missing without a trace, and it left the whole community in shock and despair. It was one of those cases that haunted everyone, especially Zoe.

"I remember feeling helpless as a child, but it is what inspired me to become a cop and try to bring closure to families who have gone through similar ordeals. It is a dark part of our town's history, but I hope to make a difference and prevent such tragedies in the future. I could not let go either. I became a detective to solve the unsolvable and get justice for people left behind."

Determination burned in her eyes, familiar to Paul's own. In sharing vulnerability, an understanding is formed between them. Their bond strengthened through complementary passions and pain, fueling their quest for truth. Zoe gazed at Paul intently. "You cannot keep punishing yourself for things outside your control. The past is the past. All that matters is what we do now."

Paul shook his head. "If I had been faster back then."

"You will drive yourself mad with what-ifs," Zoe interrupted. "Focus on the present. We are close to answers." She leaned in. "Let me help carry the burden. You do not have to face your demons alone; I am here for the long haul."

Paul looked at her, surprised by her perceptiveness. Behind the determination was compassion for his suffering. She was right. Clinging to past failures helped no one. And her support lightened the weight on his shoulders. "Thank you," he said quietly. Zoe smiled, sensing a shift. Paul exhaled, tension leaving him. For the first time, he felt ready to face the truth without fear of being consumed by ghosts. With Zoe by his side, he could confront any darkness. Together, they would unmask the evils haunting both their pasts. Paul met Zoe's steady gaze, seeing only compassion. For the first time, he felt ready to share the ghosts that had tormented him for so long.

"It was a smuggling ring we had been trying to break

up for months. Gutierrez was low level, but we knew bigger players pulled the strings." He told of the botched bust, the ambush, the crushing guilt of failure. Zoe listened intently, making no judgment. Her silent support emboldened Paul. "Now it is all connected, the money, dates, names from my case. This puppet master was there all along, and I never saw him." She sensed the closure he now dared to hope for. "You have suffered enough. Let us end this together." Her hand squeezed his arm gently. Paul smiled softly.

"Thank you. For listening and for your faith that we can do what I could not do alone." Darkness still lurked ahead, but facing it no longer filled him with dread. With Zoe at his side, he was ready to confront any demons and finally lay his ghosts to rest.

CHAPTER 6

DEAD ENDS

The morning sun rose over the Caribbean, its rays doing little to lift Paul's brooding mood. He sat in the station, poring over case files by lamplight long before the others had arrived. However, no matter which angle he studied, the details offered no new insights—the frustration of a dead end. It can certainly wear out one's spirit.

But that is part of the job, Paul reasoned. One must accept the fact that not every case will be neatly tied up with a bow. It is a reminder that in this line of work, persistence and resilience are paramount. So, despite the disappointment, Paul would say the feeling is a mix of frustration and determination to keep pushing forward, no matter how elusive the answers may seem. Zoe arrived with coffee, seeing the tired frustration in his eyes.

"Still at it?"

Whenever Zoe sees Paul after he has been working

long hours and he is frustrated, she cannot help but feel a mix of concern and admiration—a concern because she does not like seeing him in that state, knowing how much he cares about solving the case. And admiration because even in the face of frustration, he does not give up. He is constantly pushing forward, searching for that breakthrough. She tries her best to support him, offer a fresh perspective, and remind him that they are in this together. It is not an easy job, but seeing Paul's determination inspires her to keep going and never lose hope.

Paul nodded, shoving papers away, "Every thread we pull just unravels into nothing. Whoever did this covered their tracks well."

Zoe surveyed the disarray, picking up a sheet. "What about Grant? That phone log shows suspicious calls the night of the murder."

"I checked the records. Grant was having a very public dinner across town." Paul rubbed his eyes, doubt creeping in. "At this rate, Thompson's killer will get away with it."

While Paul lost himself in files from years past, Zoe

took the lead with the investigation. She decided to start with Thompson's colleagues, hoping one may reveal a hidden motive. She drove to Thompson's office and found Greg Adams, one of Thompson's directors. He gave the same rehearsed statements as before, but Zoe saw sweat on his brow.

She prodded, "What aren't you telling me?" Adams stalked off without answering. When he stormed off like that, Zoe was taken aback. She could not believe he just walked away without answering any questions. It ticked her off, but she knew it would not get her anywhere if she chased after him at that moment. Sometimes, you must let people cool down and regroup. It gave Zoe time to think about her next move and gather more evidence. She figured they would cross paths again soon enough so they could have a proper conversation.

Leaving the Thompson Building, Zoe next tracked down tracked down Alexa Chavez, Thompson's former mistress. Zoe telephoned Chavez, and they agreed to meet at a local coffee shop in one hour.

They met at Brew Haven, a downtown coffee shop that had a familiar and inviting atmosphere, perfect for a casual meeting or a moment of relaxation. The walls were adorned with artwork from local artists, giving it a vibrant and creative feel. As soon as you entered, the aroma of freshly brewed coffee wafted through the air, captivating your senses. The soft lighting created a warm ambiance, and the gentle hum of conversation provided soothing background noise. The space was adorned with comfortable seating arrangements, from plush couches to wooden tables and chairs.

Behind the counter, baristas skillfully crafted intricate latte art, creating beautiful designs atop steaming cups of coffee. The menu boasted a wide selection of drinks, ranging from classic espresso to creative concoctions with names that reflected the shop's playful spirit. The café had large windows that allowed natural light to flood in, giving a glimpse of the vibrant streets outside. It was the kind of place where people gathered, whether alone with a book in hand or meeting friends for a chat. Zoe was surprised that Chavez chose such a public place.

From the moment they started talking, Zoe could sense that Alexa Chavez had something to hide. Alexa was quite guarded in her responses, evading most of Zoe's questions and deflecting the conversation to more trivial matters. It was clear she did not want to reveal too much. But as a detective, Zoe knew she had to push a little harder to get to the truth. She tried to build a connection with Alexa, finding common ground and using charm to put her at ease.

Slowly, Alexa started to open up, revealing tiny pieces of information that could be relevant to the case. However, there was still an air of mystery surrounding her. It was as if there were layers upon layers to her story, and Zoe was determined to unravel it all. Zoe left the meeting, not learning anything new, as meeting with Alexa was a delicate dance of trust and suspicion. It was clear she had something to hide, and Zoe was going to make it her mission to find out what it was.

Zoe questioned other suspects on her list, with much the same results; Frankie Morales seemed genuinely shaken, and Maxwell Grant gave slippery answers between thinly veiled threats. Neither cracked, but her intuition told her they knew more than they let on.

It was getting late, so Zoe returned to the station in hopes Paul was still there. Paul was still sitting at the desk that he was using when Zoe left. She reported her findings as they were, hoping Paul gained a new perspective. But his focus remained inward, chasing memories instead of leads. She tried masking her frustration, but the dead-eyed stares of the suspects weighed on her as heavily as Paul's distraction. Their biggest break so far had dissolved, leaving them as lost as when they began.

Paul emerged from his fog to find Zoe poring over maps with red threads crisscrossing the city. "Another dead theory?" he asked gently.

She sighed. "Every angle leads nowhere. Who is this guy, Houdini?"

Taking the seat beside her, Paul laid a hand on the tangled mess. "This case has us chasing our own tails. Let us pull back and look at what is right in front of us with fresh eyes."

Zoe frowned. "Like what? We know it all already."

Paul gathered the files. "Indulge me, take another pass. Look for bits we glossed over, details others deemed insignificant. New eyes find new meaning in old words."

Reluctantly, Zoe took her stack. "If you say so. But I still think…"

"Just look," he urged. "See what emerges." As they read in silence, Paul felt Zoe gradually calm and focused. Hours passed unnoticed. Then, a soft gasp, "I think I see..." She pointed at witness statements, recounting an errand the night of the murder. Paul read keenly. A sliver of a new lead or another false dawn? Only time would tell if Zoe's renewed diligence bore fruit. But her enthusiasm was kindled anew, and for now, that was enough.

As night fell once more, Paul and Zoe continued working under dim lamplight. The stagnant air was thick with fatigue and fading hope. Zoe rubbed her eyes. "Greene's alibi never sat right with me. We should bring him in again."

Paul frowned. "One interrogation solved nothing. We need proof, not wild theories."

She tossed her pen. "And waiting has? At least I am trying things instead of brooding over dead cases!"

Her outburst surprised them both into silence. Paul's eyes darkened. "You think I am not trying?"

Zoe backpedaled, but tension remained. "I only mean... we are stuck, and time slips by."

Paul rose, temper flaring. "Then, by all means, go

interrogating without cause. See how far that gets us when Greene's lawyers tear our case to shreds. Experience says patience yields far more…"

"Experience stalled you for a decade!" Zoe shot back.

"Sometimes bold action finds answers where caution fails." Her challenge hung heavy as they stared off, frustrations boiling over. For the first time, cracks showed in their partnership; each convinced the other threatened the case's fragile progress. In the ensuing silence, resolve wavered, but the question remained if their differing views could still unite to find Thompson's killer.

Despite their clash, work continued in stilted silence for a few more hours; then Paul said he was going home. Without another word, he got up and left, leaving Zoe sitting alone.

Once home, Paul continued to research his past cases while Zoe pored over witness accounts at the station. One name kept cropping up: Tyler Greene.

Later that night, a revelation struck Zoe like lightning. She had to tell Paul. She thought about calling him, but he might not answer. She decided to drive to his place instead. She arrived at Paul's only to find his house dark and empty. A note pinned to the door said that he would be in the field, following a new lead alone. Zoe's jaw tightened. She was both worried and frustrated. Worried because Paul was taking unnecessary risks, especially considering his past and the dangers we had already encountered during the Thompson case.

Frustrated because she wanted to be involved in every aspect of the investigation, and him going off without informing her made Zoe feel like he did not fully trust her or her abilities. But in the midst of those emotions, Zoe also understood Paul's perspective. He had his reasons for wanting to pursue the lead on his own, probably fueled by his own experience and the need to protect me from potential harm. She knew they each had their own strengths and ways of approaching the case, and she had to trust that he knew what he was doing, even if it felt exclusionary at the time. Ultimately, it made Zoe even more determined to prove herself and show Paul that she could handle the pressure and contribute meaningfully to the investigation.

While there was a mix of worry and frustration, it

fueled her resolve to excel in her role as a detective and prove her worth. But would it cost her? She retreated to her car, pulled out Greene's file, and drove off into the night.

When Paul arrived at the marina docks, his lead had vanished. Cursing his wasted effort, he turned for home, only to spot Zoe's car idling near the old warehouse district. Heart in his throat, Paul raced over and hauled her from the vehicle.

"Are you mad, coming here alone? These men are dangerous!"

Zoe yanked free indignantly. "I was handling it fine till you barged in. Sometimes partners must trust each other's judgment."

Paul shook his head, anger battling fear. "Not when bad judgment gets you killed. Follow me out of here; we are leaving together. This reckless chase ends tonight before it claims your life or our case. Follow me, and no detours!"

His warning was clear: the next time, he might not arrive on time. By the time they reached the station, anger had boiled over. Zoe rounded on Paul. "You think I am some rookie who can't handle herself?"

He sighed wearily. "It is not about skill. It is about

experience. You charge into danger without considering the consequence."

She jabbed a finger at his chest. "And you? You are so obsessed with dead cases and ghosts that you let Thompson's killer slip away!"

Paul's eyes flashed. "I am trying to learn from past mistakes. You rush ahead half-cocked and foolishly risk your life instead of using that big brain of yours."

Ouch, her expression hardened. "All I hear is arrogance; the great detective Paul knows best as always. Well, not everyone needs your precious experience to do their job, old man. Some of us can think for ourselves!"

With that venomous parting shot hanging between them, Zoe stormed from the station into the flickering night. Paul sank into his chair, drained by their clash but still unwavering in his belief that experience was all that separated her from lethal mistakes. The rift created seemed impossible to mend as long as their views remained irreconcilable. It was a moment, Zoe rationalized to herself, fueled by frustration and a bit of impulsive annoyance when she called Paul an "old man." They had been working together on the Thompson case and had their fair share of disagreements and clashes in investigative styles. Paul's experience and wisdom were undeniable, but at times, Zoe felt like he doubted her capabilities and tried to protect her a little too much. It was not her finest moment, and she certainly regretted it later.

She knew it was impulsive and, in a way, disrespectful. But it was also a way for her to express her frustration and assert herself as a capable detective, eager to prove herself and make a mark in the field. In the end, though, she understood that Paul's intentions were rooted in concern and a desire to keep her safe. They both needed a reminder of the need for better communication and mutual respect in our partnership.

Still seething, Zoe threw herself into the case with renewed vigor the next day. She hit the streets alone, chasing down every whispered lead without telling Paul. An informant pointed her toward the docks that evening.

Again, the docks. Keeping a safe distance this time, Zoe watched two burly men haul crates onto a ship under cover of darkness. When money changed hands, she moved in, flashing her badge. The men bolted, but Zoe gave chase, sprinting over wet planks as thunder rumbled in the distance. She recognized Greene as one of the men. But suddenly, her quarry turned and ran. With frustration fueling her at losing her only leads, Zoe returned to the station, wondering if she should tell Paul.

Back at the station, she poured through records linking Greene to the smuggling ring. A buried loan agreement tied it all back to Thompson. Zoe smiled, triumphant. Let Paul say she could not get the job done without him. As clouds burst open above, Zoe worked into the night alone, energized yet exhausted. She refused to let anything slip through the cracks again, not even consulting her prickly partner. The case would be closed, with or without Paul's say-so or input. She would prove to them both what she could accomplish on her own stubborn merit.

Three long days passed in which Paul heard nothing from his elusive partner. He could not help but worry about her safety, especially considering the dangerous nature of their investigation. Thoughts of her well-being consumed his mind, and he questioned whether he had made the right decision involving her in this twisted case. He replayed every conversation, every piece of evidence in his mind, desperately searching for any potential leads that could bring her back to his side. Uncertainty and concern weighed heavily on him, fueling his determination to find the truth.

When Zoe finally stumbled into the station after yet another fruitless night, exhaustion had worn away her ire. Paul took one look and steered her gently towards a chair. "When was the last time you slept?"

She mumbled incoherently, rubbing gritty eyes. "Every lead petered out like the rest. Another failure to add to the pile."

Sighing, Paul piled case files in front of her sleepy form. "I may have been stubborn... but this case still demands our combined skills and experience. Where do we stand?"

Zoe laid it all out with stoic brevity, swallowing her bruised pride. When silence fell, she glanced up hesitantly. "Thoughts? Guidance?"

A faint smile crossed Paul's lips. "Get some rest first. Then we start fresh. I trust your judgment as much as my own. Two heads are better, as they say."

Color returned to Zoe's wan cheeks at the small

reconciliation. A new dawn and renewed partnership felt within reach if they committed to seeing eye-to-eye. With cautious optimism, they parted ways to seek rest and perspective before returning to face the mysteries, waiting with vigor rekindled.

Reenergized after a night's rest, Paul and Zoe reconvened to compare notes. A contemplative silence fell until Paul spoke. "I owe you an apology. My focus on ghosts led me astray from the present case. Old failures closed my mind when it should have stayed open."

Zoe listened without judgment. "We all have pasts that haunt us. What matters is not dwelling there alone."

Paul met her steady gaze gratefully. "You are right. I let trauma cloud my judgment when I should have had faith in our partnership. Can you find it in you to trust me again?"

She placed a comforting hand on his arm. "We are in this together, to the end. No one faces darkness alone, not you and not me. As long as we keep communicating, our perspectives will only strengthen each other."

A look of relief passed over Paul's features. "Then let's get back to work as a united team once more. Between our skills and support, this case won't stay unsolved much longer."

Smiling, Zoe gathered her notes. "Lead the way, my friend. Wherever the evidence takes us, I will have your back every step." With their bond reforged through openness and understanding, Paul and Zoe continued with renewed purpose, ready to conquer any challenge if they stood together.

With a newfound solidarity, Paul and Zoe got back to work with vigor. They mapped connections between Greene's smuggling ring and Thompson's dubious dealings, finding a web of graft that permeated local leadership. As they compiled evidence late into the night, fatigue dissolved beneath the surge of progress. A few final pieces fell neatly into place, sparking an epiphany. Paul leaned back, eyes bright. "I think I know who our killer is."

Zoe stared. "Who? How? Tell me!" He laid it all out cogently, weaving motive from the mounting clues. Zoe saw the truth in his words and grinned, slapping a congratulatory hand on his back.

"Damn, you have still got it!"

Chuckling, Paul gathered their work with satisfaction. At long last, after many dead ends, their persistence had paid off. Now came the next challenge, following the final thread to capture a murderer before he slipped once more from their grasp. As they prepared to mobilize in the morning, a new calm had settled over the partners. They had survived doubts and overcome rifts to reach this point as a united, formidable force. Come what may, Paul and Zoe now felt ready to face the depths of darkness together to claim justice's light.

CHAPTER 7

BEHIND CLOSED DOORS

It was late afternoon when Zoe noticed the irregularity in Frankie Morales' alibi. His statement claimed he was at a local bar from 9 PM to midnight, but the bartender recalled serving him much earlier.

That evening, under the cover of darkness, Paul and Zoe stealthily approached Morales' seaside cottage. The property was empty, the lights off, suggesting their target was away. As Paul and Zoe approached the cottage, a gentle sea breeze carried the scent of saltwater through the air, instantly bringing back memories of Paul's childhood by the shore.

The cottage itself was nestled among towering palm trees, creating a sense of seclusion and peace. Its weathered exterior spoke of years spent facing the elements, and the quaint porch welcomed them with creaking floorboards as they stepped inside.

"He is hiding something," Zoe whispered. Paul nodded in agreement. They ventured inside through an unlocked back door. Zoe crept in excitedly, but Paul held her back, shining his penlight around methodically.

"We need to ensure our presence goes undetected," he whispered.

They wore booties to avoid leaving footprints in the dust. Once inside, Paul examined each surface carefully under ultraviolet light in the living room. Backlighting revealed indents on the couch cushion that could match a struggle. The cottage was sparsely furnished but immaculately clean, as if recently sanitized, although it exuded a rustic charm. The walls were adorned with seashells and faded coastal artwork, a testament to Frankie's love for the ocean. The soft, natural light filtered in through worn curtains, casting a warm glow over the cozy space.

In the living room, well-worn furniture is arranged in a comfortable and unpretentious manner. The walls were lined with bookshelves, filled to the brim with an eclectic mix of novels, travel guides, and seafaring tales. The kitchen, while not particularly spacious, was neat and well-organized. The subtle aroma of freshly brewed coffee wafted from a nearby countertop, mingling with the scent of sea air. It was a place that felt lived-in and loved, where homemade meals and heartfelt conversations had surely taken place.

As they moved through the cottage, Paul discovered a small study. It was a sanctuary of sorts, with a sturdy wooden desk tucked into a corner, stacked with papers and folders. A worn leather armchair invited reflection, and the walls were adorned with maps, both old and new. It was clear that Frankie was a man of intellect and curiosity.

Meanwhile, Zoe entered the bedroom. Though simple, it held a certain tranquility. A large, open window allowed the sea breeze to caress the room, carrying the serene sound of crashing waves. A worn journal lay by the bedside, filled with scribbled thoughts and dreams that spoke to Frankie's innermost desires.

Overall, Frankie Morales' seaside cottage embodied a sense of comfort and serenity, a retreat from the chaos of the world. It was a place where Frankie had found relief in the rhythm of the ocean, and Paul could not help but feel a certain kinship with him in that serene seaside haven. Zoe, finding nothing of value in the bedroom, joined Paul in the study. Looking through desk drawers, Zoe discovered financial records listing money transfers to offshore accounts in Morales' name.

"Blackmail payments," she murmured.

Paul, meanwhile, found fingerprints that he later matched to an unsolved burglary under the desk. A muffled sound came from upstairs. Guns drawn, they crept up slowly. A flickering light shone from the master bedroom. They edged to the doorframe, held their breath, and then burst in together. But the room was empty, the TV displaying a paused video of ocean waves. Returning downstairs, Paul noticed marks in the dust as if something had recently been moved. He tapped the wall panels until one sounded hollow. Running his fingers along the wall, he discovered scraped paint and pushed, and a hidden door swung open.

"Jackpot," he said, finding documents behind it. While Paul found the hidden documents, Zoe continued her search. Inside the tiny room, Zoe's light picked out strands of dark hair adhered to the wall. Her pulse quickened until Paul pointed out cobwebs and discoloration. The space had clearly gone undisturbed for years. Still, it gave them hope that other clues remained undiscovered. They continued working methodically but swiftly.

The basement door creaked as Zoe pulled on the dangling cord. Dark and dank, they descended the rickety steps downward. Dusty shelves held jars of preserves while a workbench collected cobwebs. Then Zoe shone her light onto a locked trunk under the stairs. One pick and it swung open, revealing wads of cash amid faded polaroids. A photograph of Thompson caught her eye; he was bound and gagged, eyes pleading with his captor, Morales, who sneered above him.

"We've got him!" Zoe whispered excitedly. But Paul held up his hand. Rummaging further, he drew out a velvet bag that chinked heavily when shaken. Inside lay expensive watches, one engraved with the victim's initials.

"This is damning evidence," Paul acknowledged. "But we need more corroboration before making arrests. If Morales is our man, he will slip up again. Let's keep watching quietly for now."

Zoe was excited when she found the evidence. It felt like they were finally getting somewhere. But when Paul decided to hold off on making an arrest and keep watching Morales instead, Zoe became frustrated. She understood his reasoning, wanting to gather more information and potentially apprehend any accomplices, but part of her just wanted to see justice served right away. She held back, telling Paul her feelings about his decision, as she reasoned it is all about finding that balance between patience and urgency in a case like this.

They re-locked the trunk and left things undisturbed, slipping away into the night with hopes their patience would soon pay off. The car ride from Morales' house back to the police station was, at best, intense. At least it was for Zoe. The car they were in was an unmarked police vehicle, simple, inconspicuous, and old. It had that distinct scent of leather and a few random pieces of paperwork scattered on the seats. The interior was a bit cramped, but it did not matter at that moment because Zoe's focus was solely on the evidence they had just uncovered and what it meant for the case. The tension in the car was palpable as they discussed their next moves and prepared themselves mentally for what lay ahead.

Filled with anticipation and a sense of urgency, Zoe still wanted to arrest Morales, but Paul simply said, "Not now."

Back at the station, Paul rushed DNA samples from

the crime scene through the lab. Hairs found at Morales' cottage were a partial match for Thompson. But without a direct sample from their suspect, the results remained circumstantial. Zoe pored over records of Morales' activities, noticing increasing vitriol and threats towards Thompson in recent phone logs. But hard proof of motive was still absent.

Days passed as they kept Morales under close watch. Paul grew tense seeing their target slip past the net again. Then a call came; Morales' boat had been spotted leaving port under cover of darkness.

Paul and Zoe raced to the docks, finding the slip empty. On the ground near Morales' boat slip lay cigarette butts, recently crushed. DNA swabs were later matched to the partial profiles from the crime scene. Still, at least in Paul's mind, without directly catching Morales in the act, their case was not airtight. Frustrated but keeping their cool, they agreed stealth was their best tactic for now. The hunt would continue until their suspect made a fatal mistake.

Back at the station, Paul loaded the photos from Morales' cottage onto his computer. His eyes scanned figures and account numbers documented in the perp's rushed scrawl. He double-clicked to enlarge a bank statement. The listed account bore only a number, but it triggered Paul's memory.

"Take a look at this," he said to Zoe. She leaned in, brows furrowed as Paul overlaid the statement with records from Ray Thompson's offshore activities. The account numbers matched.

"So, Morales and our victim were both using the same overseas banker," Zoe murmured. "Maybe to launder ill-gotten gains?"

Paul nodded slowly. "Could be blackmail, could be business. But it gives us firmer proof of a connection. I say we pay Mr. Chen another visit."

Zoe smiled. "Let's rattle some cages and see what falls out."

Their renewed sense of momentum drove them to pursue every angle. Each tiny clue nudged them nearer to the dark heart of the mystery shrouding Thompson's demise. Zoe leaned forward eagerly. "I will contact the bank in the Caymans right away."

Again, Paul stopped Zoe. He shook his head. "Let's tip our hand until we are ready to play the ace. One wrong move could sabotage the case."

She sighed in frustration. "But every moment we delay, more of their money gets laundered."

He placed a calming hand on her arm. "I know you are impatient; it is what makes you such a firecracker of a detective. But we must build an airtight case before confronting these financial kingpins. Subtlety is our ally; we will watch and wait until all the players are in position, then sweep the board clean in one fell swoop."

Zoe took a deep breath and released it slowly. "Alright, we do this your way. Just promise me we will not let the grass grow under our feet."

Paul smiled. "Not to worry, we will pursue our leads with relentless precision. Soon, all the twisting threads will join together, and the total picture will come into focus. Justice will be served, I assure you. For now, patience remains our sharp-edged tool."

Her frown turned to a look of thoughtful resolve. "Then let the countdown begin."

That evening, Paul and Zoe were still poring over files long after the station emptied. Paul traced web-like patterns between the perps with a red string on the evidence board.

"Look here, Morales' records show regular payments to this shell company, which we know funnels to our banker friend. And over here, Thompson received similar monthly deposits. I would wager those originated from the very same place."

Zoe nodded slowly, absorbing the connections. "You're right. It is all starting to come together now. I would never have seen the full picture without your guidance."

Paul acknowledged her praise with a humble smile. "Together, step by step, the smallest clues fall into devastating alignment."

As the moon rose, Zoe fetched strong coffee to fuel their pursuit of the elusive truth. Working as a seamless team, the darkness began to peel away. By late night, the true scope of corruption was illuminated in grim relief. Exhausted but satisfied, they concluded their work with a renewed sense of closeness.

Zoe gazed at Paul with new eyes, grateful for his steadfast mentorship. With him at her side, any mysteries seemed solvable.

With work over for the day, Paul treated Zoe to a late meal. The late-night break with Paul was a much-needed moment of respite for Zoe, especially during a tough investigation. Paul took her to a quaint little deli tucked away in a quiet corner of the city. It was not anything fancy or upscale, but it had that cozy, welcoming atmosphere. The walls were adorned with old photographs and vintage posters, giving the place a nostalgic charm. They sat at a small table near the window, sharing their sandwich choices and enjoying a momentary break from the intensity of the case. It may not have been a high-end restaurant, but the simple yet delicious sandwiches and the company made it a memorable experience.

As they worked on their hearty sandwiches and coffee, the conversation turned personal. "You are diligent, maybe to a fault,"

Paul noted. "Do not lose yourself so deeply you cannot find your way back to daylight hours. This job will chew you up if you let it dominate your every waking moment."

Zoe ducked her head, chastened. "It is just so absorbing, you know? We are trying to bring justice to these people, but their lives still matter when we clock off. I want to give it my all."

Paul nodded understandingly. "Commitment is admirable, but self-care is crucial too. Make time for family, friends, and hobbies. Recharge your spirit so you can show up as your best self every day. I am always here if you need an extra pair of eyes or someone to talk to. But promise me you will strive for balance, yeah? A happy detective is a sharp detective."

She smiled, reassured. "Thanks, Paul, for looking out for me both professionally and personally. I will try to heed your wisdom. Now, tell me about what you do to unwind. I could use some relaxation tips!"

Chuckling, Paul began regaling her with stories of island life, providing levity amid their serious work. But also some not-so-lighthearted stories. He shared a few captivating stories from his days as a detective. He shared tales of some of his most puzzling cases, recounting the intricate details and the twists and turns that ultimately led to the resolution.

One story involved a missing person's case that took him across multiple cities, while another revolved around a high-stakes heist that seemed like something out of a movie. Paul's storytelling skills were impeccable, and he had a way of drawing Zoe into the narratives, making her feel like she was right there with him, experiencing the excitement and challenges firsthand. It was a fascinating glimpse into his past adventures and a chance for Zoe to learn from his wealth of experience. After storytelling, sandwiches, and coffee, both left strengthened in body and spirit to face tomorrow's unknown challenges. As they prepared to part ways for the night, Paul mulled over their findings.

"We have gathered compelling links between Morales, Thompson, and their questionable finances. But is he truly our killer, or merely a piece meant to lead us astray?"

"I have been wondering the same," Zoe admitted. "His involvement seems too obvious. But we cannot ignore the evidence right under our noses."

Paul nodded slowly. "Keep an open mind. Stay vigilant. Our real prey may still lurk in the shadows. For now, we have made progress in unraveling this twisted tale. With care and intuition to guide us, the final secrets will not stay buried."

Zoe sighed. "Another long day of work ahead. But with you at my side, I feel ready to face whatever this case throws at us next." She bade Paul goodnight, determined to give her all to the investigation. Though doubts lingered on the identity of Thompson's true killer, their partnership blossomed with each new discovery. Justice would be theirs in time.

CHAPTER 8

DANGERS EMERGE

Paul and Zoe returned to the crime scene in the late afternoon the next day. Paul had a nagging feeling that they might have missed something vital during their initial investigation. Going back was necessary to ensure they had not overlooked any crucial evidence. As for what they were looking for, Paul still had his eye on a particular suspect, Morales. There were some inconsistencies in his alibi that he wanted to examine further. The goal was to gather more evidence to build a strong case against him. Paul knew they needed to be absolutely certain before making any arrests.

Zoe was eager to go back. She had a palpable sense of excitement as they reentered the crime scene. She saw it as another opportunity to prove herself and make progress in the investigation. While their experiences and perspectives differed, Paul appreciated her enthusiasm and drive to solve the case. It is important, Paul knows, to have a partner who shares the same level of commitment.

Paul admitted to himself, without sharing his thoughts with Zoe, that he was feeling a mix of frustration and determination. Frustration because the case seemed to be hitting a dead end, and determination because he knew we could not afford to give up. As for when they would apprehend Morales, as Zoe wanted to do, Paul insisted it depends on what they find during their investigation. Paul kept telling Zoe they must gather enough concrete evidence to prove Morales' involvement beyond a reasonable doubt. Rushing into an arrest without a solid case will not serve justice.

"Rest assured," Paul told Zoe before they arrived at the crime scene, "We will not rest until the truth is uncovered and the guilty party is held accountable."

While Paul was outside looking for clues, Zoe remained inside the crime scene, meticulously examining everything she saw. She combed through the surroundings, not overlooking anything, regardless of how small or insignificant it was. Her focus was on collecting any potential leads or overlooked details that could help solve the case.

While conducting her search, she thought again

about trying to persuade Paul to arrest Morales. It did not take her long to realize she should not. They had discussed Morales as a potential suspect earlier, and they both agreed to keep an open mind and gather more evidence before making any conclusions. Zoe had relented and finally agreed they should be thorough and ensure they had a solid case before taking any action. She had agreed, but that does not mean she was happy about agreeing.

Paul was combing the grounds of the crime scene, scanning for any clues the detectives may have missed on previous visits. A glint in the sand near the edge of the property caught his eye. He crouched down and brushed away the grains to find an indented boot print partially covered. Frowning, he studied the print. The depth and detail were too clear; this print was fresh, not weeks old like it should have been if left at the time of the murder.

A sense of unease crept up his spine. He stood and surveyed the area more closely. Nearby, he found scuffed bushes and torn branches on the otherwise neatly trimmed hedges. Someone had been here recently after the police had wrapped up initial investigations and deemed the scene secure. But why? What were they looking for? Paul's frown deepened as numerous worrying possibilities sprang to mind. He returned to Zoe, who had wrapped up the interior search of the house, and met Paul by the driveway.

"We have a problem," he said grimly. "It looks like someone paid a return visit to our crime scene after we finished processing it. I think we are being watched, and now I am wondering how much the killer has been able to monitor our investigation."

Zoe's eyes widened in alarm. She glanced back at the roped-off property, suddenly feeling exposed under unseen surveillance. "But that is impossible... how could they have gotten past security?"

"I don't know," Paul replied. "But I intend to find out. In the meantime, we need to be extra careful. This killer has been one step ahead of us the whole time. We cannot afford to make any mistakes from now on."

After leaving the Morales house, Zoe and Paul went their separate ways. Paul thought it necessary for them to cover more ground and gather additional evidence individually. Zoe headed back to the station to review some case files and cross-reference them with the information we had gathered so far. She wanted to dig deeper into the connections between Ray Thompson and the unsolved crimes from Paul's past cases.

Paul decided to revisit some of the locations related to Thompson's business dealings and his connections. He wanted to look for any potential witnesses or any overlooked details that could shed light on Thompson's murder. Paul spent the afternoon visiting various establishments, talking to employees, and trying to piece together the puzzle. Their temporary separation allowed them to gather different perspectives on the case. They were both driven by the same goal of uncovering the truth and solving the mystery, so they both knew it was essential to use their individual strengths to their advantage.

Zoe drove along the deserted coastal road, the sun setting hours ago, leaving her vehicle as the only source of light on the empty stretch of road. She yawned, rubbing her tired eyes. It had been a draining week with few leads and even fewer breaks in the case. But she knew they were close; she could feel it in her bones.

A shape up ahead caught her eye. She slowed, peering through the darkness. But it was just a fallen tree branch blown into the road by the sea breeze. Shaking her head at her overactive imagination, she sped up again. That is when she saw the pair of headlights rapidly gaining in her rearview mirror. They closed the distance between their vehicles too fast, barreling upright on her bumper. She eased off the gas, intending to give them room to pass. But the other car remained hard on her tail, blinding her with high beams. When she tapped the brakes in a hint, they only accelerated closer.

Fear gripped her heart. She slammed down on the accelerator, swerving as the car rammed into her back fender. The abrupt turn threw her into a fishtailing skid. Gripping the wheel in white knuckles, she fought to regain control. Just as her tires found purchase again, a dark shape loomed ahead where the road should be empty. She wrenched the wheel right, narrowly dodging an instant head-on collision by inches. Her pursuer was gone as quickly as they had appeared, leaving her unharmed but shaken on the shoulder. Someone did not want their investigation to continue...

After a few minutes to regain her composure, she continued her drive home. By the time Zoe got home, she was still shaken by this incident. She had to calm down, somehow. She took some deep breaths to steady herself. She has always found that deep breathing helps her relax and clear her mind. Then, she made herself a hot cup of herbal tea, one of her go-to stress relievers. Finally, she sat down in her favorite chair, closed her eyes, and tried to focus on the present moment. She tried to let go of the thoughts and emotions related to the accident and allowed herself to unwind. She tried. She thought about listening to soothing music or doing a quick meditation to calm her nerves further but decided she was too tired and just wanted to go to bed. Sleep seemed like the best self-care at this point.

Just as she was about to fall asleep, glass shattered loudly inside Zoe's apartment. She jumped up with a gasp as a brick crashed through the window, sending shards spraying across the room. Heart pounding, she edged over gingerly and nudged the brick with her toe. A folded piece of paper fluttered out, bearing a cryptic message: "Drop the case or else." Zoe shivered, gripping her arms. She did not scare easily, but this was personal, and the implications were chilling. Too many thoughts were swirling around in her head: Should I call for backup? Should I install security cameras? Where is my gun? But before she could fully process the attack, pounding footsteps sounded outside.

"Zoe? Open up!" Paul called urgently. She yanked the door open to reveal him wide-eyed and battle-ready, having sprinted over at the first sound of breaking glass. His gaze raked her frantically for injuries before settling on the threat, still lying on her carpet. "Are you alright?" His hands skimmed her arms, throat, and shoulders as if checking for wounds himself. Satisfied she was unharmed, he turned steely. "This ends now. I will not let them hurt you."

Zoe looked at Paul and asked, "How did you get here? Why are you here?"

"I had just arrived after checking on a possible lead nearby. It was just by chance coincidence that I happened to be pulling up when I heard the sound of breaking glass."

When the brick was thrown through her window, Paul heard the sound, and he acted swiftly to ensure Zoe's well-being. Fortunately, Zoe was not harmed, but it was a stark reminder that the killer was willing to go to great lengths to protect their secrets. His words both soothed and fired Zoe's determination. Let the killer try threatening her;

she had Paul now, and they had, and will, overcome anything to expose the truth. This case was too important to back down, no matter the cost.

Paul called the station and had two officers respond and watch Zoe's house from outside the rest of the night, much to Zoe's dissatisfaction. Zoe prowled her house restlessly, going mad under the watchful eyes of the officers stationed outside her door for her "protection," as if she needed babysitting. The threats only fueled her drive to solve this case - she didn't want or need coddling.

"They are just trying to keep you safe," Paul reminded gently. "We have upset someone with our progress, so we have to be careful."

Zoe sighed, knowing he was right. Still, being sidelined went against her nature. "I should be out there helping you investigate, not cooped up in here."

"And I should be tracking leads, not playing bodyguard." Paul's mouth quirked wryly. "But someone tried to run you off the road tonight and threw a brick through your window. Until we know more, this is necessary." Zoe grumbled but did not argue further. At least

Paul understood her frustration, even if he did not share it. His presence was a small comfort despite the circumstances. She just hoped this extra protection paid off and did not slow their momentum toward solving the case. The killer would not stop their threats easily, and she was itching to go on the offensive.

Paul stayed with Zoe that night, sleeping on the couch. Neither had a full, restful sleep, but enough sleep would keep them going the next day. The next morning, while drinking coffee in Zoe's kitchen, Paul's cell phone rang. It was Harry, one of Paul's former informants. Harry is a former associate of Ray Thompson and claims to have some valuable information regarding the case.

After hanging up, Paul said, "I have known Harry for quite some time, having crossed paths with him during my previous investigations." Paul explained that Harry has a knack for being in the know, in the right place at the right time, and gathering tidbits of information on various individuals in his line of work. Though his methods may not always align with the law, Harry has proven himself to be a useful source of information in the past. However, like any informant, Paul cautioned Zoe that they must approach

Harry and his tips with caution and verify their accuracy through our own investigation.

Their protection detail had refused to greenlight this meeting, but they felt it offered clues worth the risk. Paul and Zoe entered the rundown police informant's building on high alert. The building is a rather dilapidated structure situated on the outskirts of the city. It is far from the hustle and bustle of the nicer downtown area, which is the picture postcard of the Caribbean, surrounded by overgrown vegetation and remnants of abandoned businesses. The streets are quiet and almost forgotten, which makes them an ideal hiding place for someone like Harry, who prefers to keep a low profile.

The building itself exudes an air of neglect and decay. It is an old, three-story brick structure, its exterior weathered and stained with the passage of time. Broken and boarded-up windows are scattered along its facade, evidence of years gone by without proper maintenance. A worn-out wooden door marks the entrance, its paint peeling and hinges creaking with each movement. Inside, the hallways were dimly lit, the flickering fluorescent lights casting an eerie glow on the cracked and chipped walls. The carpeting, now threadbare and worn, adds to the sense of neglect that permeates the building. The scent of dampness and neglect lingers in the air, hinting at the lack of care the building has received over the years.

Inside, a scruffy man eyed them warily. "You the cops?" At their nod, he spat. "Ain't got nothin' to say."

Paul slid a folder of records across the table. "You recognized this man?" His eyes flickered.

"Maybe. Whaddya want?" A scuffle sounded outside. Paul whipped around just as the door crashed open. Two hulking men bounded in brandishing pistols.

"Hand over the file!" one bellowed. Paul surged into action, tackling the closer assailant. They crashed to the floor in a tangle of fists. Zoe dodged a backhand from the other, kicking out his knee. He howled but swung his gun at her. She grabbed his wrist, struggling for control. A shot rang out. Zoe froze in horror, but Paul had disarmed his man and fired up at the ceiling.

"Enough!" The informant bolted out back. Paul cuffed the thugs as sirens wailed in the distance. Zoe's heart pounded, buzzing with adrenaline. Someone was desperate to hide behind those records... Harry's SUV screeched out of the alley, almost sideswiping Paul's car.

"Time to go," he barked, stomping the gas. Zoe grabbed her police radio. "All units, suspect vehicle fleeing east on 5th, attempt to intercept. Consider the suspects armed and extremely dangerous. Approach with caution." Paul swerved wildly through traffic in pursuit, getting in front of them in an attempt to have them stop. The SUV rammed the rear bumper, trying to spin them out. He downshifted, acceleration shooting them forward enough to pull alongside.

Zoe saw the driver reach under his seat, and she screamed a warning. Paul jerked right as a hail of bullets tore through where they had been. Civilians scrambled out of the way as the chase escalated, sirens blaring closer. Ahead, the SUV fishtailed around a corner, straight into an ambush of police cruisers appearing from side streets. It lurched to a halt, crashing into a telephone pole, and its gas tank ruptured. Paul and Zoe leaped out, weapons drawn. The driver bolted from his door and tried to flee on foot with the envelope of evidence. Paul tackled him, and they rolled amid the dirt and debris. The crazed man, gripping a flare, lit it and tried to burn Paul.

Zoe ran to help Paul subdue him, distracting the man long enough for Paul to land a blow, knocking the flare away. It bounced, spreading paper ignition across puddles of fuel. Sensing doom, they hauled the driver away moments before the site exploded behind them in a fiery ball. Shaken but intact, they cuffed their man safely at last. Paul and Zoe stared at the blazing fire, its ferocity mirroring their inner turmoil. They had come dangerously close to becoming the killer's next victims.

Paul requested the uniformed officers to secure the area and take custody of the assailant. He would call ahead and request an experienced detective conduct the interview to gather information and pass the information on to Paul. Zoe turned to Paul, her face ashen. "They tried to kill us. This is not just some ordinary criminal; he or she is a cold-blooded murderer and will do anything to avoid capture."

Paul met her eyes grimly. "I know. We rattled their cage and pushed them into a corner. From now on, we cannot make any assumptions about what they can do."

Zoe shuddered, the events of the day crashing over her. "But we are so close to the truth. If we stop now..." Her voice trailed off in frustration and something close to fear.

Paul shook his head firmly. "No. More than ever, we need to keep going. The killer has shown their hand. We rattle them because we are getting under their skin. We are threatening to expose whatever dark secret they are hiding." His dark gaze held hers with steady resolve. "And the closer we get, the more dangerous it will become. But we cannot back down now. Justice for Thompson depends on us finding the nerve to push onward."

Zoe nodded, mustering her wavering courage. "Whatever it takes, we will see this through to the end. Together." They decided to each go home, get cleaned up, and then meet in two hours at the station. However, Zoe could not wait that long. Less than an hour later, Zoe took a steadying breath before knocking on Paul's door. He opened it quickly, relief evident as he scanned her for injuries. "Are you okay?"

She nodded. "I am fine. Look, I know we said we would be more careful, but I cannot just hide out. I need to help crack this case."

Paul studied her serious face, seeing the same resolve that drove him. "I understand. But from now on we watch each other's backs, no solo moves. Agreed?"

"Agreed. The killer thinks violence will scare us off, but it is having the opposite effect."

A faint smile touched Paul's lips. "Fear or not, we stay focused on the truth. It is the only thing that will protect us now." Paul offered to drive, leaving Zoe's car at Paul's house.

As they hurried to Paul's car, Zoe asked, "Any new leads we need to follow up?"

"A name in the records, Maxwell Grant. I want to reinterview him as our prime suspect." Paul's eyes were steely.

"Let's go rattle some cages. We will go first thing in the morning." Zoe gripped her weapon, heart pounding with mingled fear and purpose. By working as a team with Paul, she believed they could overcome anything to solve this case. And finally, bring the murderer to justice.

Unbeknownst to Zoe, Paul drove them to the police target range. He set up targets at the secluded range while Zoe watched in silence. She had no idea they were going to the range. She did not balk. He watched as Zoe emptied her clip with steady precision, remembering her poised reactions under pressure that day. "Not bad, not bad," Paul said with a smile. "Now let us do some scenario training. No time for fear when your life is on the line," he said.

She nodded, jaw set. "Bring it on!"

He threw her into choking holds, testing her defensive skills. Cool under duress, she countered swiftly each time. When Paul demonstrated disarming an assailant, she mirrored his moves until they flowed as one, a seamless unit. Later, gasping on the mats after an exhausting bout, Paul smiled at her through the sweat. "You have got guts, Walker. I am lucky to have you watching my back."

She beamed, pride blazing through the fatigue. "It is an honor, sir. Learning from the best will keep us both alive to crack this case." He believed her. Zoe had grit and smarts to match any officer. Most of all, she never backed down when the killer tried forcing them to yield. Her courage proved an inspiration as they marched toward the final showdown.

CHAPTER 9
ROCKY PARTNERSHIP

Paul paced back and forth in the police station, running his hands through his hair in frustration. They were no closer to solving Thompson's murder, and all the leads had gone cold. Zoe was at her desk, going over case files yet again. "What if we re-interview Alexa Chavez?" she said. "Maybe she'll slip up under pressure."

Paul shook his head. "It's not enough. We need hard evidence, not suspicions." He slammed his palm on the table, making Zoe jump.

"I'm just trying to brainstorm," she said defensively. "Since you shot down my other five ideas."

"Because they're just wild guesses!" Paul snapped. He took a deep breath, rubbing his forehead. "I'm sorry, I didn't mean to yell. I'm just... frustrated with hitting dead ends."

Zoe stood up. "Well, maybe if you'd listen to my

theories instead of dismissing them out of hand, we'd have a breakthrough by now."

Paul sighed. "Your theories are based on intuition, not facts. We need facts to build a case."

"And sitting here getting angrier by the minute is helping how?" Zoe shot back. Their eyes locked in a tense stand-off. The fissures in their partnership had grown into full-fledged cracks. As pressure mounted with no answers in sight, it seemed their ability to work as a team was fracturing along with the case.

Zoe felt frustrated as if the dead ends they were hitting were because of Paul. They were going in circles with this case. She wanted to make progress, solve it, and prove herself as a detective. She was, once again, having mixed feelings about working with Paul. While she respected his experience and knowledge and the fact that he had been there for her, she still felt he was holding back on her, not fully invested in helping her crack this case.

After the argument with Paul, Zoe needed some space to clear her head and reflect on everything. She decided that she needed a change of scenery to help gather

her thoughts and gain a fresh perspective. Leaving Paul to himself, she took the opportunity to review the case, thinking about the potential leads they had gathered, the evidence, and any connections they might have missed. She wanted to figure out how to move forward and break through the obstacles they had encountered. Zoe decided it was time to reevaluate and find a new approach.

The next day, with nothing but files for company, Zoe's thoughts churned darker. She should have trusted Paul's experience. And now they are not talking. Doubts plagued her. What if she wasn't cut out for this job after all? Paul had the instinct and intellect to solve the hardest crimes. Zoe rubbed her tired eyes; sleep had not come last night, re-reading the same lines over and over without comprehension. She wasn't just doubting her abilities now; she was doubting her decision to even become a detective. Maybe it was time to admit defeat, hand in her badge before she screwed up again.

That evening, Zoe worked up the courage to knock on Paul's door, takeout in hand. "Can we talk?" Paul sighed but opened the door wider to let her in. As they ate, Zoe laid it all bare: her doubts, mistakes, and recklessness. "I just feel so useless now."

Paul set down his fork. "Zoe, I don't doubt your skills. But this case has already taken enough lives. I couldn't forgive myself if something happened to you too." His voice softened with concern, not condemnation.

"You're the best partner I've ever had," he continued. "But you need to trust me and my experience."

Zoe hung her head, chastised, "I'm sorry. I'll listen from now on, I promise."

Paul nodded. "And I'll try to hear your ideas without shutting them down. We're stronger as a team." He offered an olive branch, her badge. The tension eased as their partnership was repaired. As they cleaned up, Paul spoke quietly. "I try not to show it, but sometimes the pressure gets to me too. On tough cases, I'd obsess over every detail for days." Zoe glanced up, surprised. Paul always seemed so in control. "I'd look at the same files over and over, barely sleeping," he continued. "It was like I had tunnel vision, unable to see outside the case."

Zoe murmured that she understood the feeling. "How did you cope?" Paul sighed. "Made mistakes, burned out...until I learned balance." He dried his hands slowly. "This one with you feels different. Easier. Having a partner watch my back..." His words trailed off, but Zoe heard the empathy. She wasn't alone in struggling under pressure. And Paul's experience lent compassion, not criticism, for her flaws.

"We'll solve it together," she said firmly. Paul nodded, a faint smile breaking through as they renewed their focus on the case ahead.

Zoe left Paul feeling lighter as if a weight had lifted. Their talk repaired more than just their working relationship - it reconnected the bond of trust she thought lost. Paul understood the pressures in a way no one else could. And he didn't condemn her weaknesses but empathized through his own. It was cathartic to share her doubts freely, without fear of judgment. Returning to the case renewed her drive tenfold. She wasn't just helping to please her mentor anymore. They were in it together - flaws, struggles, and all. Two equals with a shared goal of justice. Tomorrow held answers; she felt certain. But more than that, it held the prospect of strengthening the partnership that gave her wings to fly.

She and Paul would see it through to the end, no matter the challenges. Their bond was stronger now, woven of shared burdens lifted and light ahead rekindled. Nothing could stop them from solving this case as the united team they were destined to be. Sleep came easy that night, her rest untroubled by doubts.

Zoe arrived early the next day, bursting with newfound energy. "Let's look at the financials again," she said. Paul smiled, glad to see her spirit restored. They dug in, poring over records side by side. After an hour, Zoe noted odd rounding in a utility payment. "See how it's off by 73 cents? That never happens by accident." Intrigued, Paul pulled up others from the same quarter: small discrepancies accumulated, a telling pattern.

"Creative accounting to hide embezzlement perhaps?"

Zoe beamed, delighted in their synced thought process. More anomalies emerged: late payment fees waived without reason and invoices with wrong dates. The details were minor, but together, they hinted at obscured financial misdeeds. Paul's experience and Zoe's fresh eyes united to catch what one alone may miss. Collaborating re-energized them both. Their bond strengthened further, minds melding into a cohesive unit. By day's end, both felt certain a breakthrough was within reach. The reforged partnership was guiding them closer to the core of this troubling case.

Acting on their discoveries, Paul and Zoe paid a visit to Thompson's accountant. Jonathan Hanlon. Hanlon's office was located on the third floor of a commercial building in the downtown section of St. Anne. It had a professional yet somewhat cluttered appearance, with stacks of papers occupying the desk and shelves filled with financial records. The walls were adorned with framed certificates and degrees, lending an air of credibility to the space. He was a middle-aged man with a receding hairline, thin-rimmed glasses, and a slightly haggard appearance. Paul did not consider him a suspect at that point in the investigation. But he couldn't rule anyone out with a close connection to Thompson. Hanlon's involvement with Thompson's financial affairs made him a person of interest.

The man squirmed under Paul's piercing gaze. "We know you've been covering tracks. Tell me what's really been going on with the books." Paul's low voice held an edge like a knife.

"I—I was just following orders, I swear!" Their source broke out in a sweat.

Zoe watched, rapt, as Paul bore down. "Whose orders? Start talking, or you'll be sharing a cell."

Details spilled out then: creative accounting to hide offshore millions, bribes to loan sharks and police. Thompson ran a full-scale criminal enterprise.

"Names. Now." Paul was a juggernaut, unstoppable in his pursuit of truth. Some of the names included business associates and clients of Ray Thompson, as well as individuals connected to Thompson's financial dealings.

But one name stuck out. Frankie Morales.

When they left, Zoe breathed, "How do you do that?"

Paul glanced over. "Fear is a powerful motivator. But kindness works too, sometimes." His gentler tone with her, compared to witnesses, revealed the dynamics of his interrogation craft. Watching him work fueled Zoe's admiration for her partner's expertise. The day dragged on—the name Frankie Morales wore on their minds. By evening, Zoe slumped despondently at her desk while Paul paced restlessly.

She rubbed tired eyes, frustration mounting. "We're never going to solve this. What's the point of staying late?"

Paul paused, eyeing her with empathy. "I know it's draining, but quitting won't bring us answers any faster." He rested a hand on her shoulder reassuringly. "Truth has a way of emerging on its own terms, not ours. All we can do is our due diligence and be ready when it comes." He gazed out the window into the inky night.

Zoe sighed. "I wish I had your patience. But you really think it'll come if we just keep going...?"

Paul nodded, smiling faintly. "I've yet to see a mystery outlast a good detective's perseverance. Come, let's grab food. Fresh eyes may see what's been missing."

Heartened, Zoe stood. "You're right. One more look can't hurt." She felt the balance returning, her partner's anchoring faith reinvigorating their shared mission. Endurance would solve this case, as Paul had solved so many before.

CHAPTER 10

STORMY SEAS

Zoe was about to make a near-fatal mistake. She went to the docks alone that night because she thought she had a lead on a potential suspect. Zoe is aware of her independent nature and sometimes a little too eager to follow a lead. She didn't want to wait for Paul or anyone else, and it turned out to be a risky move, as she ended up underestimating the danger. She wasn't sure why, but here she was. Zoe has learned to trust her instincts and be cautious in potentially dangerous situations. She knew the docks held a haunted past for Paul, and she was aware of the significance it held for him.

She couldn't help but feel a twinge of anxiety, knowing she was stepping into a place that had caused him pain in the past. It was a reminder of the darkness and danger that can lurk in the shadows. However, she knew she had to put her emotions aside and focus on the task at hand. Her priority was gathering evidence and uncovering the truth.

So, despite any reservations she had, she steeled herself and kept alert, keeping a close eye on anything that looked suspicious, as well as any potential clues or connections that might further the investigation. As she hid behind some crates nearest the warehouse she suspected Thompson of conducting business in, a hand clamped over her mouth from behind. She struggled against the strong grip to no avail.

As her vision blurred, she cursed her impulsiveness. Paul had been right, as usual. Now, it seemed she may pay the ultimate price for her recklessness. The last thing she saw before slipping into darkness was the dock lights glinting off a sharp, steel blade.

Zoe came to with a painful groan. Her hands were bound behind her back, and her mouth gagged. Shadowy figures loomed over her in the flickering lights. She thought of Paul, hoping he'd realize she was missing soon. He was the best chance she had.

Meanwhile, Paul paced the station anxiously. He should have heard from Zoe by now. Something wasn't right. He put out feelers to his contacts, combing the city. Hours passed with no word. Dawn broke with still no sign of her.

Paul was reaching the end of his tether when a call finally came. One of his old CIs thought he'd seen Zoe being bundled into a van by the docks. Paul sped to the location, reviewing blueprints in his mind.

As he stealthily approached the abandoned warehouse, he prayed he wasn't too late. Peering through a grimy window, his blood ran cold. Zoe lay bound and gagged on the floor while two figures loomed over her menacingly. He had to act fast before they made their final move. Stealth and surprise would be his only advantages in the coming confrontation, and Zoe's life depended on them.

Zoe awoke inside a gloomy warehouse, hands still bound. Her captors were arguing nearby, revealing shreds of their plan. "Told you we should've finished her," one growled.

"Patience. She may still prove useful," the other soothed.

Zoe inched along the floor, searching for anything to aid her escape. Her fingers closed around a rusty metal pipe. If she could just get free... But her movement attracted attention.

A boot connected with her ribs, knocking the wind from her lungs. Stars exploded in her vision. Her assailant loomed over her prone form. "No more games, little girl." He raised a knife, steel glinting murderously.

Then, a shot rang out, and he crumpled soundlessly to the floor. Paul stood in the doorway, gun smoking in his steady grip.

"Get away from her," he said calmly to the remaining man. The killer took a wild swing at Paul with a pipe. Frankie Morales! They grappled fiercely, a blur of fists and feet. Zoe strained helplessly against her bonds as the fight raged. Finally, Paul gained the upper hand, subduing Morales in a chokehold until he collapsed.

Rushing to Zoe, Paul quickly untied her wrists. "Are you hurt?" he asked gently, hugging her close in relief. Their ordeal was finally over. Paul surveyed the warehouse, formulating a plan. By then, Morales had regained his composure and was pacing agitatedly. Zoe watched Paul from the corner, eyes pleading.

In a sudden rush, Paul launched himself atop Morales, Morales grabbing Paul's gun. They crashed to the floor in a tangle of limbs. The weapon skittered away across the concrete. The killer fought viciously, landing heavy blows to Paul's ribcage. Warnings from past injuries flared, but adrenaline powered through the pain. Zoe cheered Paul on silently from the shadows.

Spinning deftly, Paul gained the mount. But still, the man struggled, fingers scrabbling at Paul's face. With a snarl, he flung Paul off of him like a ragdoll. They rose unsteadily, squaring off for the last stand. Frankie Morales feinted left and swung right, connecting with Paul's jaw. Stars exploded behind his eyes, but he shook it off. Paul threaded through an opening in the man's defenses, seizing his arm in an unassailable hold. Bone and tendon ground audibly under the strain. With a guttural yell, the killer submitted and fell limp. Paul rushed to free Zoe as Morales writhed on the floor, vanquished. Paul had prevailed against the darkness. Justice was won.

After apprehending Morales, Paul made sure to secure him using handcuffs, ensuring he wouldn't be able to escape. With Morales in custody, Paul called for backup and coordinated transport to the station. Paul ensured the prisoner was placed securely in the back of a police vehicle, under careful watch, and then Paul followed in his own car to maintain security. Safety, Paul believed, is paramount, especially when dealing with a dangerous suspect like Morales. Zoe refused medical attention at the scene, promising instead to stop by the hospital later.

Back at the station, Paul paced angrily while Zoe nursed her bruises guiltily. "What were you thinking?" Paul exploded. "I told you not to rush off half-cocked!"

Zoe winced, face burning, chest and ribs aching, "I'm sorry, I just thought... if I could confirm one thing..."

"One thing could've gotten you killed!" Paul growled. "This isn't a game, Zoe. These criminals are ruthless and will do anything to cover their tracks. We have had this conversation several times!" Paul did not hide his frustration.

She nodded miserably. "I know, I wasn't thinking straight. It won't happen again, I promise."

Paul dragged a hand down his face in frustration. "I can't keep bailing you out. Next time, you might not be so lucky."

Zoe met his gaze, eyes brimming with shame and

regret. "I know, and I'm grateful you came for me. Thank you, Paul. You were right; I should've waited."

Seeing her remorse, Paul's anger deflated. He sighed wearily. "Just promise me no more solo missions, okay? We're stronger as a team."

Zoe smiled gratefully. "I promise. Partners from now on."

Paul nodded, tension seeping from his frame. Another crisis was weathered through their trust in each other. "I'm sorry for letting my frustration get the better of me earlier," Paul said with a sigh. "We're both under a lot of pressure here."

Zoe shook her head understandingly. "No, you were right to call me out. I could've jeopardized everything." She extended a hand to Paul. "From now on, we do this together, as a team."

Paul grasped her hand firmly. "Together. No more rushing off half-cocked or letting our egos get in the way. We trust each other." A comfortable silence fell as they got back to work, tired bodies running on determination alone. This case tested them both physically and mentally. But through the trials, their partnership had only grown stronger, and they finally got their man.

As a retired detective with vast interrogation experience, Paul took the lead in questioning Morales. With Morales securely detained, Paul began the interview, employing his usual methodical approach aimed at extracting crucial information to uncover the truth behind the crimes. Despite his persistent questioning, Morales remained tight-lipped and, at times, confrontational. Paul sought to unravel the intricate web of deceit surrounding him but made little headway. It was a tense and grueling process, with Morales attempting to throw Paul off track with evasive answers and false leads. The interview had its share of challenges, but ultimately, the arrest of Morales was a pivotal turning point in the case, leading them closer to the truth and justice. A breakthrough came late into the night, a connection between Morales and another unsolved homicide.

Excitedly, Paul and Zoe pieced together the new clues. Could this finally be their big break? Renewed vigor filled them. Though darkness yet lingered, their bond gave them hope. If they stood united against the shadows, justice would prevail. This was a fight they were resolved to win side by side.

With teamwork and trust, Paul and Zoe ventured into the final confrontation, ready to defeat evil once and for all. The end was approaching, and victory would be theirs. Paul stared into space, lost in memories, only half-remembered. "It never gets easier, watching someone you care about put themselves in danger," he said quietly. "I've lost partners before, good men and women who gave their lives trying to fix this broken world. And it nearly destroyed me each time."

Zoe listened silently, sensing there was more he needed to say. Paul sighed. "The last one... it was my fault. I should've had her back, but I missed the signs and underestimated our prey. Pledged it would never happen again." He met Zoe's eyes, vulnerability laid bare. "That's why I get so overprotective. The thought of losing someone else..."

Reaching out, Zoe squeezed his hand comfortingly. "We're in this together now. You don't have to carry the weight alone anymore."

A ghost of a smile crossed Paul's face. "I know. And

I'm grateful... for your partnership. It gives me hope that this time, we walk away victors."

Their bond was sealed in that moment of shared sorrow and strengthened determination—two fractured souls, healing as one under the night's watchful stars.

Zoe rummaged through her pockets triumphantly, "Look what I managed to sneak out of the warehouse," she said, showing Paul a crumpled receipt and a strand of fibers.

Paul examined the items closely. "You risked your life for this?" His voice was stern, but his eyes betrayed concern.

She nodded. "I knew it was dangerous, but I felt I had to try. These could be the break we needed."

Smoothing the receipt, Paul's eyes widened. "This purchase places our suspect at the scene of the crime. The fibers also match a distinctive rug only found in one place." The clues fit like pieces of a puzzle.

After so much danger and loss, hope flared anew that justice may yet prevail. Zoe watched Paul connect the dots, relief washing over her. Though the road was long, their partnership paved the way. With skills and trust combined, the reward of truth seemed finally within their grasp. All that remained was the final confrontation to end the darkness once and for all. Paul smiled at Zoe, pride shining through past worry. "You did well. Now let's finish this." Hand in hand, they raced towards the climax with justice as their guiding light.

Paul took a steadying breath as the fateful address came into view. After so long spent running, it was time to face the dark truths of his past. Beside him, Zoe squeezed his arm supportively. "We go in together, just like always. Whatever demons dwell within, we'll conquer them side by side."

He nodded gratefully. "Your partnership gives me strength. With you at my back, I feel ready to take on any shadow from my past or present. Our bond will guide us to victory."

Standing before the door, they shared a look, steeling their resolve. No evil would break their unity on this day. As one, they kicked in the door and strode boldly into the lair of the killer, coming to terms with ghosts long buried. Paul faced his trauma with Zoe by his side, illuminating the dark corners his mind had locked away. With her support, the memories no longer held power. He was free.

United, no foe could deter their mission to bring the criminal to justice. Through every trial, their teamwork would triumph. Together, they were unstoppable. This day would end the nightmare, once and for all.

CHAPTER 11

LIGHT IN THE DARK

Zoe and Paul crept through the dark warehouse, guns drawn, silently cursing her impulsiveness. Paul's warnings echoed in her mind, but she was determined to prove herself capable. After entering, Zoe went left, and Paul went right. Despite the haunting memories for both Paul and Zoe, they had to push through their fears and confront the demons of that warehouse once again. It was a test of resilience, forcing them to face the past to uncover the twisted truths of the present. It was an uneasy and unsettling experience for both.

Returning to the warehouse, however, was necessary to uncover the truth and connect the dots in the current case. Paul felt the air heavy with anticipation, for he knew that within these walls lay the keys to unraveling the twisted truth. He was a tangled mess of hope and trepidation, driven by an unyielding determination to confront the ghosts of the past and unmask the killer's motives.

Suddenly, a noise in the shadows drew her

flashlight's beam to a figure wrapped in a tarp: Alexa Chavez! Zoe's gasp turned to terror as rough hands grabbed her from behind. She struggled vainly as a voice whispered, "Naughty girl shouldn't snoop." A prick in her neck brought darkness.

As Paul heard the commotion, he approached, and then someone struck him across the back of the head, causing him to become unconscious. When she woke, Zoe found herself bound, eyes finding Paul's gaze across the room, acting as though he was still out. Together, they watched their captor pace, talking to himself about his plans to eliminate witnesses. As his back turned, Paul met Zoe's eyes and nodded subtly. With frantic wriggles, she freed a hand to retrieve her knife from her pocket and flicked it open. When the man turned back, Paul lunged, distracting him from the blade slicing through Zoe's bindings.

As sirens neared, their joined tackle brought their foe crashing down in a tangle of limbs. Breaths heaving with reaction, they shared a smile of partnership renewed through peril confronted side by side. Zoe shivered in the dim light, calculating her chances.

Footsteps approached as the door creaked, but relief

flooded her at Paul's fierce scowl. "We end this now," he vowed. As the door opened, a scream went up in the air. They charged forward to find their foe strangling a bound man. In the ensuing brawl, Zoe's well-aimed kick freed their ally. Seizing the chance, Paul grabbed their choking foe, Tyler Green. With a tight smile, he said, "Justice will be served, I promise you both."

The tension broke as help arrived at last. Zoe caught Paul's weary gaze, gratitude, and partnership strengthened in peril faced with foes defeated. Morales and Green were both in custody.

At the station, she watched Tyler Green being hauled away to a cell, leaning on Paul with gratitude for his timely rescue. Though the case ended, their bond as partners pursuing justice was only beginning. Zoe pored over evidence with renewed focus, ignoring exhaustion. A connection between victims stirred the recollection of a clue. She led Paul eagerly to her desk. They pieced the puzzle together, now understanding, unseen pieces falling into a grim pattern. Where others saw defeat, Paul had sensed the truth's nearness.

Zoe grasped now why his guidance proved

invaluable, experience guiding her growing instincts. Re-examining leads under this new light revealed long-hidden meaning, motivation emerging in clear lines of malice. As dawn broke their vigil, the victory felt nearby. Paul's steady counsel found an unwavering ally in Zoe's perseverance. United in justice's call, darkness, every deception stood illuminated before their eyes. Thanks flowed between them as partners, faith strengthening their strides down the final path to resolution. Justice would see its day through their efforts, and a new day its dawning as well.

Green sneered at their questions, armored by legalisms. But facing Paul unleashed more; his dark gaze promised this man's ruin if truths remained buried. Cracks split the veneer as he leaned near, murmuring of lives left shattered. Zoe smiled watching the suspect cave, appreciating Paul's tactful coercion unveiling each clue. Paul laid out every piece of evidence linking Morales and Green to the murders of Thompson and Alexa Chavez. Every piece of evidence links them to money laundering and smuggling—every piece of evidence links them to Garcia.

Each discovery buoyed fresh determination until, at last, the tide could be turned, exposing long-hid culprits and clearing clouded waters for justice to flow clear once more. Strands stretched from names, tracing amounts and dates, weaving patterns through the years with ruthless logic. Corruption's roots had grown deep, but nourished darkness could not stand against clarity's searchlight.

During the tense interrogation, Paul asked Green why Thompson and Chavez were murdered. Green revealed that Thompson was murdered because he had become a liability to certain individuals involved in illegal activities. According to Green, Thompson had discovered information that could expose their criminal operations and had threatened to go to the authorities. As a result, they saw fit to silence him permanently.

As for Chavez, Green claimed that Chavez had to die because she had discovered something that threatened to expose Green's illegal activities. It seems Chavez stumbled upon some incriminating evidence, and Green believed that eliminating her was the only way to protect their secrets. As flickers fused into beams, the facts emerged from its smokescreen for Zoe and Paul to comprehend at last.

Where before all was cloudy, now all stood illuminated as if dawn seized midnight's stolen robes. Facts once disparate now merged as waves designing their separateness to a clear direction, now cresting towards shores where retribution might ground. Paul and Zoe emerged from the interrogation, exhausted but satisfied. Each new revelation strengthened Paul's case, drawing tight around those who thought themselves beyond the reach of justice's retribution.

Now, all stood exposed to the clarity Zoe and Paul brought, tearing away every artifice by dint of perseverance and the paired reason's light, which illuminated long-buried bones. Thanks flowed between them for the triumph of truth emerging after so many fruitless days. Zoe watched knowledge kindle in Paul's eyes as past financial phantoms assumed solid form. Threads trailing from years past re-wove their pattern anew before his sight. Where cynics saw mazes twisting endlessly, his faith and experience discerned design.

Zoe saw now why experience had called Paul from retirement, his wisdom guiding her from their first meeting, a meeting that was formed by puzzles, to this clarity. Gratitude flowed between them for victory hard-won through perseverance against corruption's hydra. Though more quests doubtless awaited, this conquest proved their faith well-placed.

Purpose renewed, they turned toward the new day thus unveiled. Zoe's excitement met Paul's caution as closure drew near. Hard-won clarity buoyed hopeful plans yet sobered by memories haunting her mentor. Together, they grasped victory and the toll it exacted, wisdom tempering youth's exuberance. Her smile thanked his guidance, and her esteem was growing where once had stood only doubt. Bonds born of strife held fast though somberness veiled them, now each understanding the other's demons masking behind purpose's noble mask.

For Paul, there was a surge of mixed emotions coursing through his veins. Relief, indeed, but also a profound sense of satisfaction. It was a triumph, a culmination of tireless investigative work and the unraveling of countless webs of deception. However, that feeling was tempered by a sobering realization that justice alone cannot undo the past or erase the pain caused by the killer's actions. Yet, at that moment, as Paul stood face to face with those responsible, there was a flicker of satisfaction in knowing that he had prevailed and was one step closer to closing the chapter on this troubled case.

After capturing both suspects, Zoe definitely had a sense of accomplishment. She felt like it was a major breakthrough and a step towards solving this tangled case. However, she couldn't help but feel a lingering sense of unease. Something told her that there was more to this story, that there were still unanswered questions.

As for thinking the case was finally over, she really wanted to believe it. They had worked so hard and put in so much effort. But deep down, she knew there was more to uncover. There were still loose ends and unanswered mysteries that needed to be resolved.

With all the said feelings and emotions between Paul and Zoe regarding this case, they didn't really have a chance to share a quiet moment of congratulations. They were both focused on wrapping up the case and finding the truth that they didn't take the time to celebrate prematurely. There was also that nagging feeling that there was still more to be discovered. They still had work to do.

CHAPTER 12
CONFRONTING GHOSTS

There was still one more suspect Paul wanted, Guiterrez. The heavy rain pelted against the windshield as Paul and Zoe sat in silence, surveilling the home of Guiterrez through their binoculars. The rain created a constant drumming sound. The wipers worked furiously to clear the glass, but the rain kept coming down relentlessly. The air inside the car became humid and a little stuffy despite the coolness outside. Zoe found herself constantly shifting in her seat, trying to get comfortable, and occasionally glancing at Paul, wondering how he seemed so unfazed by it all.

To pass the time, they chatted in hushed whispers, going over their plan and keeping their eyes glued to the house. But as the hours wore on, Zoe could feel the monotony settling in like a heavy fog. It was a delicate balance between staying alert and fighting off the temptation to succumb to the drowsiness that threatened to creep in. Strangely, this kind of stakeout in the rainstorm added an extra layer of tension to the whole situation. The sound of raindrops and the dimly lit surroundings only heightened the suspense.

Surprisingly, Guiterrez's home was a modest-looking two-story residence nestled within a quiet neighborhood. The exterior was painted a muted shade of blue, with a neatly trimmed lawn and a couple of potted plants flanking the entrance. Nothing too extravagant or eye-catching, which made it easy for Paul and Zoe to blend into the surroundings as they sat in a car across the street.

Little could be seen in the darkness, but they knew he was hiding something crucial inside that would lock away his fate. A light flickered on in an upstairs window. Zoe nudged Paul and handed him the binoculars, her breath fogging the glass. Through the rain-spattered lenses, Paul saw the killer pacing back and forth, clutching files to his chest. When he turned, the lamp caught his profile; it was unmistakably him.

"That's got to be where he's keeping the evidence," Paul whispered.

Catching movement by the door, they ducked down

in their seats. But it was only a shadow, not the killer emerging.

Zoe tapped her foot impatiently. "We need to get in there and have a look before he destroys it all."

Paul nodded slowly, gaze still fixed on the window. Lightning flashed, outlining the killer's harried movements. "Wait for the right moment. He'll slip up eventually..." As midnight struck, the killer's shadow crossed the window one final time before his light flickered out. "Now's our chance," Paul decided.

They darted across the lawn, sheltering beneath the porch from the raging storm. Paul picked the lock swiftly. Inside, all was dark, eerily still. They surveyed the empty rooms, rain beating against the windows.

"He's left in a hurry," Zoe said. A glint caught her eye; she scooped a crumpled note from the debris and read, "Meet me at the docks, midnight." Paul's blood ran cold.

Remembering their last two visits to the docks, he said, "It's a trap." But he couldn't take the chance. They had to go. They raced to the dock as lightning cracked—the docks at night. Paul saw the docks as a place filled with an eerie sense of mystery and uncertainty. The darkness engulfs everything, making it difficult to discern any details or potential threats that may be lurking nearby. Each step is accompanied by the echoing sounds of water lapping against the old wooden piers and the distant creaking of ships.

The air carries a salty tang, a reminder of the vast ocean beyond. In this darkness, Paul thought, your senses become heightened; every creak, rustle, and gust of wind prompts a cautious glance over the shoulder. Your heart beats a little faster, aware that danger could be hiding in the shadows. The only sources of light are the dim flickering bulbs lining the docks, casting long, foreboding shadows that seem to dance and taunt. There's an unshakeable feeling of vulnerability, knowing that at any moment, an adversary could emerge from the depths of the night. The sound of footsteps, distant voices, or even the echo of your own breath seems amplified, causing you to strain to listen for any signs of trouble.

But amid the tension, there's also a sense of camaraderie and determination as Zoe and Paul navigate this dark and uncertain terrain. Paul knew they had to rely on each other, their shared purpose guiding them through obscurity. It's a constant battle between the desire to solve the case and the nagging thoughts of safety. Yet, driven by the need for justice, they pressed forward, persevering in pursuit of the truth, even in the face of such uncertainty. A dark shape appeared from the darkness, the killer, untying a rope on a boat tied to the pier.

"It's over!" Paul shouted above the thunder. But the man only smiled, striking a match in his cupped hands. A boom shattered the sky, flames mushrooming upward. Paul hurled Zoe into the water as wooden shrapnel rained down. From the depths, she watched the boat consumed; two black shapes writhed atop the burning wreckage until all was silence but the screams of the whirling storm.

Coughing up smoke, Paul dragged Zoe from the water as the dock collapsed. "Are you alright?" He rasped, drawing his gun. Movement, the killer fled into the stormy night. They chased him into dark, empty warehouses and alleys, now flooded with rain, the killer's tracks fading in puddles lit by intermittent flashes of lightning.

Exhausted, Paul and Zoe paused to catch their breath. Then, a bolt of lightning illuminated a figure watching from a building ahead. The killer, taunting them to follow. They advanced cautiously through the skeleton frames rising against the roiling sky. The killer ran into a tall, dark, and mostly empty building.

"Nowhere left to run!" Paul yelled. The man merely smiled, footsteps echoing as he continued into the dark building. Zoe moved to a side door to cut off his escape. Another flash and he was gone. They rushed into the darkness of the building, hearts in mouths—chasing the man up several flights of stairs to the top.

Again, Paul yelled, "Give it up, you are trapped!" Then, unexpectedly, the man smiled then jumped. A shape twisted unnaturally on the rubble five floors below. The chase was over, but would justice be served? Only the thrashing clouds knew for sure. From the killer's twisted form rose a ghastly chuckle, quickly swelling to a wild cackle over the storm's howl. Paul and Zoe peered down in horror, transfixed by the crazed laughter.

"Well done, Phillips! It seems you've finally caught me," the man called up. "Only a little too late, as always." Paul's blood ran cold at the voice, one that had haunted his dreams for over a decade.

"Gutierrez," he growled. "It was you behind it all."

"Bravo for putting it together... at last!" Gutierrez mocked. "That botched bust in the shipping yard really scared you, didn't it? I enjoyed every moment; the look on your face as you struggled was priceless."

Fury and disgust rose in Paul's throat as old failures and new realizations crashed together. Zoe grasped his arm, anchoring him in the present even as ghosts assailed from the past. For now, justice has remained just out of reach. But answers were coming soon, one way or another... Gutierrez cackled madly, his twisted delight muffled by the raging storm. "Did you enjoy watching me torment you all these years?" he called up to Paul. "So close, yet never close enough."

"Why?" Paul demanded, struggling to contain his fury. "Thompson, my old case, what was the point?" A savage grin split Gutierrez's ruined face.

"Collins. Remember Collins, your old partner?" he sneered, "He was getting too close to my operation. I couldn't let him expose me, so I put him in your care. When you failed, I knew you'd haunt yourself forever. As for

Thompson, he wouldn't pay his dues. But his money helped fund so many... projects."

Zoe recoiled in horror and rage, all the lives destroyed by this one man's evil machination. "It ends tonight," she vowed.

Gutierrez threw back his head and laughed. "You think it's really over, detective? I have powerful friends, eyes everywhere. You'll never be free of me." With that, his manic cackling faded into the howling wind, cruel mirth echoing even in death.

At long last, Paul had his answers and his closure. But at what cost? Grief and anger consumed Paul at the revelation. With a wordless cry, he ran down the stairs to Gutierrez's body below. But as he descended, a shadow emerged from behind a concrete pillar below—another follower, equally crazed. They grappled as the storm raged, the man shrieking his devotion to Gutierrez's vision.

The man who had emerged from the shadows to attack Paul and Zoe was a tall and imposing figure, towering over them with an air of menace and sending a chill down Paul's spine. His build was muscular and formidable, indicating physical strength that posed a serious threat. His face was weathered, etched with lines and scars, a testament to a life lived in darkness and turmoil. His eyes gleamed with a cold, calculating intensity devoid of empathy or mercy. The slight stubble on his jawline betrayed his negligence towards personal grooming, suggesting a man consumed by his single-minded pursuit.

Clad in dark clothing that blended into the night, he moved with fluidity and precision, exhibiting a dangerous level of agility. Every movement seemed purposeful and deliberate, as if he had honed his skills to be a silent predator in the shadows.

His demeanor exuded a palpable sense of aggression, as if violence was his default response to any perceived threat. Paul thought there was a certain familiarity in his eyes, a glimmer that hinted at a shared history or a past connection that Paul couldn't quite place. In that intense moment of confrontation, fear mingled with determination within Paul. It was clear that this man was willing to go to great lengths to protect his secrets and escape the long arm of justice.

But Zoe and Paul stood firm, ready to defend

themselves and unveil the truth, even against such formidable opposition. Paul was slowly overwhelmed by the madman's strength. Zoe raced down, struggling to get a clear shot as they thrashed in the half-light. With an animal snarl, the man threw Paul aside, seizing Zoe's weapon in one swift motion.

"No one will stop us now!" he howled, pressing the muzzle to her temple. Paul could only watch in anguish, battered and unarmed. All seemed lost beneath the shroud of night and rain.

But then, a sudden crack of thunder masked Zoe's swift maneuver. Her elbow crashed into the man's gut, and they tumbled, the gun skittering away into the dark. Lightning flashed on the steel as Zoe dived for salvation, praying she was not too late...

As Zoe fought for the gun, memories engulfed Paul, visions of past failures he'd spent years trying to forget. Faces of the lost swarmed his mind's eye: Collins, bound and bleeding; innocent lives cut short due to his mistakes. Rage and sorrow merged into a roar that tore from his throat, drowning out the thunder. He flew at the man, grappling ferociously despite torn flesh and broken bones. This time, he would not fail.

Through the storm, Paul battered his opponent, each blow avenging those who'd suffered before. Lightning threw their savage shadows across the scaffolding in a primitive, primal dance of death. At last, the man faltered, toppling from Paul's unrelenting onslaught. As Zoe trained her weapon on their adversary, Paul stood poised to end this nightmare once and for all. But through the sheets of rain, he saw only ghosts pleading for release. With a yell, he thrust his hands down, pressing the man's face into the concrete tiles until the skull met unyielding certainty. Only then did the phantoms recede, allowing the storm to slowly wash the blood from his hands along with the demons of his past. It was over.

As the storm began to subside, a familiar mad cackle echoed through the construction site. Panic rose in Paul's throat as another shadow emerged through the gloom. The real killer, Guiterrez, triggered by their disruption to dismantle his empire, was going to try and end this once and for all. In his hands, a detonator primed to level the place and all within.

Paul knew they had mere moments before

obliteration. With Zoe's help, he rushed to disarm the bombs planted through the scaffolding as the killer closed in, howling with glee.

A tense game of cat and mouse ensued through the skeletal frames, each side racing against time. At last, in a flash of lightning, Paul tore the detonator from grasping fingers, tossing it high above. Guiterrez screamed his denial at the roiling clouds. Then thunder answered, and the world exploded in a blinding fireball. When the smoke cleared, all that remained was a smoking crater... and two souls free at last from the chains of the past.

Exhausted but victorious, Paul took Zoe's hand and limped slowly into the fading storm, the ghosts of failure blown away on the retreating winds of change. A new dawn was rising over a city with lives reborn. The morning sun broke through swirling clouds, its light finding Paul and Zoe huddled amid the rain-soaked rubble. Eyes met, reflecting a new dawn of understanding and acceptance. Paul smiled wearily. "It's finished," he said. "At long last, the ghosts can rest."

Zoe squeezed his hand and helped him stand. "You did it," she said proudly. "You faced your demons and won."

They surveyed the charred ruins where so much evil had festered. Now, only ashes remained, blown clean by the storm's final howls into a bright new day.

"Let's go home," Paul replied. One last call to make first. He radioed Dispatch, calmly detailing Gutierrez's decade-spanning crimes and their bittersweet resolution. Justice would be served for Collins, Thompson, and all the nameless victims. As sirens wailed in the distance, Paul exhaled deeply and let the past slip away on the breeze. Hand in hand with Zoe, he turned to face the future and whatever challenges held next for this partnership, now and always equal, whole again.

The case was closed. A new phase began.

CHAPTER 13
TRUTH UNMASKED

Despite Paul's detailed presentation of evidence meticulously pieced together over the course of the investigation, Morales and Green remained defiant, refusing to admit culpability. Meeting Paul's gaze with cold indifference, they dismissively waved away each damning fact as mere "coincidence" or "circumstance."

"Luckily for society, the court of law does not operate on your word alone," Green sneered. "It will take more than conjecture and twisted logic to convict me."

Unfazed, Paul merely arched an eyebrow. "That may be so. Fortunately, hard facts don't require belief or faith to stand as irrefutable truth." He gestured to the evidence board behind him. "Each piece links to the next in a clear chain too unbroken to deny. But you're right that the law will have the final say. And when it does, justice will be served."

With that, he gathered his files, ready to present an

airtight case that left no room for doubt or misinterpretation. Their arrogance would soon be shattered by cold, harsh reality. For a while, denial offered a feeble defense. The truth was immutable. And Paul held it in the palm of his hand. Settled across from Green and Morales in the interrogation room, Paul began calmly detailing his reconstruction of events.

"On the night of Thompson's murder, cell phone records place you, Green, at the scene. Financial records show regular payments from Thompson into your personal accounts, suggesting blackmail. And this encrypted drive contains damning evidence of past criminal conspiracies you both took part in."

Green growled in frustration but said nothing. "That same night, eyewitnesses saw your vehicle parked near where Thompson's body was found. And footage from a nearby security camera captures two figures, matching both of your descriptions, disposing of his corpse. Forensic analysis matched soil samples from the crime scene to your boots, Green, and your shoes, Morales. Fiber evidence tied you to the ligature marks on Thompson's wrists. And your prints were lifted from the murder weapon." Paul pressed on, painting a damning picture move by deliberate move, stripped of doubt or plausible deniability.

His evidence dissected motives, means, and opportunity with chilling precision. By the time he wrapped up the hour-long presentation, Green and Morales knew their fates were sealed. Paul had proven beyond question their direct roles in both past conspiracies and Thompson's and Chavez's murders. Justice would be done on this day in more ways than one.

Leaning forward, Paul fixed on Green and Morales with an intense stare. "You thought you got away with it for so long. Pulling the strings from the shadows while others did your dirty work. But not this time." A nerve twitched in Morales' jaw.

"You have nothing."

"Don't I?" Paul mused. "I've reconstructed every move. Decade old crimes and new. All leading right back to both of you and Guiterrez."

"Coincidence!"

"Is it?" Paul tilted his head, scrutinizing Morales' cracking facade. "Or were Thompson's threats to go to the police with what he knew just too much of a risk?"

"You can't prove a thing," Green said, obviously stressed. His expression gave him away, eyes flitting as memories resurfaced of that fateful night.

Paul had seen that tell before, the subtle signs of guilt emerging under pressure. And so, he applied more, chipping away ruthlessly at the killer's fraying composure until, at last, it all came crashing down in a torrent of incriminating words.

"I had no choice!" Green burst out. "He was going to destroy everything I built!"

Paul sat back with a small, satisfied smile. And with that, the case was closed before it ever began. The truth had won out in the end.

From the observation room, Zoe looked on in awe as Paul methodically broke down their carefully constructed walls of deceit. Utilizing both hard evidence and calculated psychological tactics, he peeled back the layers to reveal the true extent of their depraved plans. It began as simple blackmail and bribery; Morales was forced to admit under Paul's stern but unyielding interrogation. But the lust for power and wealth grew until it consumed him. No one was beyond their reach or influence. Politicians, police, and judges all danced on his twisted strings.

"And Thompson," Paul pressed, "what was to be his fate if he refused to stay bought?"

"Death," Green seethed, "just as it was for the others. Loose ends had to be eliminated, no matter the cost."

Zoe shuddered to think of the untold victims whose fates hung in the balance of these men's monstrous greed and arrogance. But Paul was unflagging, peeling back layer after sordid layer with relentless precision. By the time his interrogation ended, the corruption had been laid bare for all to see.

In Paul, Zoe saw a guardian of justice, wielding both facts and innate humanity to drag hidden evil, kicking and screaming into the light. Her awe of his mastery knew no bounds.

While the confession secured their fate, Paul was not so easily satisfied. Justice had been served for Thompson and the other past victims, but the threat remained. This one puppet master controlled an entire web of corruption;

severing the head did not kill the hydra. As they exited the interrogation room, Zoe sensed Paul's lingering focus. A part of him was still in there, piecing together loose threads.

"It's over now," she ventured. "You got them, and Guiterrez is dead."

Paul nodded slowly. "I got them. But their influence spread wide and deep. Dismantling a network takes more than an open-and-shut case."

Zoe was a bit disheartened that Paul did not accept the case's resolution. However, she understood. If any tender remained, the dark designs could yet live on through other hands. "What do you need?"

A grim determination set in Paul's eyes. "Names. Connections. Everything we tore out of them about others on the web. With what we know now, we follow the string to its outermost reaches. We rip it all out, root and stem, and scatter it to the four winds. Only then would the threat be ended, the slate fully wiped clean of this twisting evil." And so, their work continued, chasing shadows now into the light of justice's piercing gaze.

The following morning, news of a predawn convoy

ambush brought Paul and Zoe racing to the scene. Officers were down, the armored vehicle crippled, and its prisoners freed in a hail of gunfire. As the sun began to stir from its slumber, casting a faint glow on the horizon, a group of assailants orchestrated a well-planned and daring attack on the prisoner transport vehicle. The scene was shrouded in darkness, the only source of light being the dim streetlights scattered along the road.

The silence of the early morning was shattered by the sudden eruption of gunfire and the screeching of tires. The sharp staccato of bullets echoed through the air, mingling with the shouts and cries of both the attackers and the transport vehicle's security personnel. The assailants moved with precision and coordination, their faces hidden behind masks, striking with calculated ferocity. They unleashed a storm of bullets upon the transport vehicle, aiming to disable it and liberate their imprisoned comrades. The transport vehicle's security personnel valiantly fought back, taking cover behind the vehicle while returning fire.

Amid the chaos, panic, and adrenaline, time seemed to stretch and warp. Each second felt like an eternity as the fate of those riding in the transport vehicle hung in the balance. The air was thick with tension and uncertainty, the outcome of the ambush teetering on a knife's edge. The scene was illuminated by sporadic muzzle flashes, casting fleeting glimpses of the combatants locked in a deadly dance. Sirens wailed in the distance, growing louder as reinforcements were en route, but time was of the essence. The clash of will and determination continued until one side gained the upper hand.

The attackers, satisfied with their successful extraction, vanished into the dawn's fading darkness, leaving behind a trail of death, destruction, and unanswered questions. The ambush on the prisoner transport vehicle was a stark reminder of the lengths that some would go to protect their own or to disrupt the wheels of justice. It was a tense and disorienting moment that demanded split-second decisions and risked precious lives. Only through unwavering determination, skill, and a bit of luck could those caught in this deadly encounter hope to emerge unscathed.

Paul examined the wreckage with a thoughtful eye. "This was no random act but a precision strike to free their prized assets. Their accomplices wanted loose ends tied up for good by any means necessary."

Zoe voiced the grim realization. "They'll be on the run soon."

But Paul shook his head. "Not if we get to them first."

From the evidence, he had deduced their likely safehouse hideout. If he was right, they had a short window to intercept the killer on his journey into the netherworld. In their examination of phone records and surveillance footage, Paul noticed a pattern of frequent calls and messages originating from a specific phone number. This number appeared to be a central communication hub coordinating the activities of the assailants. By triangulating the signal tower data, they managed to narrow down the general vicinity from where these signals were emanating.

During Paul and Zoe's search of the abandoned warehouse used as a temporary hideout by the assailants, they discovered a handwritten note containing partial addresses and coded phrases. Using Paul's years of experience, he deciphered the code and realized that it referred to specific landmarks and street names within a certain area of the city. Combining these significant pieces of evidence, along with his intuition and experience, they deduced that the likely hideout of the assailants could be found within a rundown neighborhood near the waterfront. The close proximity of the area to the initial crime scene and the analysis of their communication patterns pointed Paul toward this conclusion. With this newfound information, they prepared to venture into the heart of darkness, guided by a determination to bring the culprits to justice. Sirens blaring, they sped off into the morning. Zoe admired Paul's cunning, outmaneuvering darkness at its own game yet again.

When they crashed through the ramshackle doors, weapons drawn on a startled gang, their quarry was indeed in tow. "Leaving so soon?" Paul's voice carried an edge like steel. Victory was nigh. One final showdown remained to eliminate the root of this far-reaching evil once and for all.

With practiced efficiency, Paul and Zoe disarmed the gang members. But their true target stood guarded by two familiar faces, Green and Morales, eyes burning with fury. "You thought you could outrun justice," Paul addressed the killer calmly. "But I'm always one step ahead. Your accomplices' efforts were futile; this ends now, today, right here."

The killer barked a vicious curse but made no other reply. His companions, seeing the trap sprung, moved to attack. Unfortunately for them, Paul was more than prepared for such dramatics. A flashbang exploded, blinding the group in its glare. When senses returned, both enforcers lay groaning in cuffs while Paul stood unchanged, the killer clutched before him. "It's over," he whispered in the killer's ear. "Your legacy of poison dies with you, its foul works undone. I've followed every thread back to its root, snipped each tender from the stem."

Today, the sun shone its brightest yet. Paul smiled faintly as his colleagues around him swarmed to take charge of the sealed case, whose solving cemented his place as a guardian of justice's light.

Carlos Diaz's violent reign had reached its end at long last. With Diaz thrown screaming into the prison transport, along with Morales and Green, the long ordeal drew to its victorious close.

Zoe turned to her mentor, eyes shining. "You did it, Paul. After all this time, that monster will never hurt another soul. Justice is truly served tonight because of you," Zoe beamed.

A small, satisfied smile lifted Paul's weary features. "It was never just about one case or criminal. We did this. I could not have done it without your help, young detective. I couldn't have torn down his whole web alone."

No praise was needed; being part of such an achievement was reward enough for Zoe. To have worked alongside a master, learning skills to safeguard others - made all the hardship worthwhile. Paul sensed her pride and placed a fatherly hand on her shoulder. "Now, the future is yours to shape. But should darkness arise again, do not forget where there is injustice; good people will rise as well. As long as hope remains, evil can never prevail. Take that lesson with you always."

Zoe nodded, eyes bright with purpose. A new dawn had risen indeed.

While justice's victory here was assured, Paul saw one loose thread remaining that gave him pause. "This isn't over yet," he murmured to Zoe as the transport pulled away. "We have the confession and evidence aplenty, but the trial remains. And where there's a will for corruption, influence finds a way."

Zoe frowned. "You think their connections may challenge the outcome?"

"It's a risk," Paul admitted grimly. "We tore out the roots, but poisonous spores could remain. I want airtight security on the case files. And we should brace the judge and jury discreetly, make sure every precaution is in place."

Zoe took heart as well, Paul's vigilance leaving no avenue unexplored. "Have faith," she told him. "We'll be watching. No trickery or intimidation will stand a chance against the truth. Their day in court will be their end."

Paul hoped she was right. Only the trial's end would lift all doubt. But with allies like Zoe at his side, even that bitterest battle might still be won.

Back at the station, a sleepy-eyed but jubilant team congratulated the triumphant detectives. Paul and Zoe accepted smiles and handshakes warmly but without abandoning their focused demeanor.

"The mission isn't fully complete," Paul reminded them quietly.

Despite the killers being in custody, their evil yet swayed unseen in the dark. With Diaz gone, those strings had lost their puppeteer but still posed risks if left uncut. He needed their help following the final leads, severing every last tie to darkness.

Zoe backed the request, appealing to their shared commitment to see justice served wholly. The team understood. Though tired, purpose hardened their resolve. They would keep digging until no shadow of this twisted web remained. And so, renewed in goal if not body, the hunt continued into the night.

Paul glanced at Zoe with a small smile, pride swelling within. She had grown into a superb detective, guiding others as adeptly as she once followed. Their partnership, tested in the crucible, now blazed twice as bright against the enveloping dark. Whatever evils still lurked, this light would not waver or fail until all shadows fled before the dawn.

CHAPTER 14
ONE LAST STAND

The day of Diaz's trial loomed over Paul and Zoe like a dark storm cloud. Green and Morales had already been convicted and sentenced. Neither would see the light of day outside of the prison walls.

As they reviewed the case files one last time, news arrived of another attempt to sabotage justice. Two witnesses had recanted their testimonies overnight, and another suffered a mysterious "accident." It was clear Diaz's allies were aggressively trying to seed doubt and cover tracks. Paul examined the witness statements closely, noticing subtle inconsistencies that revealed intimidation. He deduced the true threats lurking beneath polite rescinds of information. Zoe saw the same and realized just how dangerously powerful the killer's network still was, even from behind bars. They redoubled security for the remaining witnesses and evidence. Paul suspected bribery was also at play and had the financials of key figures monitored closely. Sure enough, unexplained deposits appeared, tracing back to slush funds controlled by Diaz's advisors on the outside. It seemed the trial would be an uphill battle against a well-oiled machine of corruption. But Paul and Zoe were determined to withstand every underhanded tactic, preserving enough integrity in the justice system to secure conviction. The web had not yet been fully dismantled it seemed. The final showdown was only just beginning.

In the days leading up to the trial, Paul and Zoe, along with the district attorney, toiled non-stop, preparing their evidence. They reinforced weak points, fortified strong ones, and considered every angle of attack the defense could take. Despite their exhaustive work, the killer's allies continued finding ways to cripple their case. Another investigator was bribed into transferring crucial records offshore, disappearing them into a legal black hole. Explosive witness testimony was threatened into submission through allusions to exposing past indiscretions.

Zoe felt her rigorous preparations starting to slip as the sands shifted beneath her feet. Paul sensed her unease and redoubled his own efforts, determined not to let their oppressors gain an inch. But even his renowned focus showed cracks when an apparent accident landed a colleague in the hospital. The implication was clear: no one was safe until the trial concluded.

As the opening arguments drew near, Zoe found Paul pacing restlessly, worrying at doubts like a dog with a bone. She saw the toll the constant threats were taking and knew they had to find a way to wrest back control or risk losing everything they'd fought for at the trial. The web was tightening, and they were starting to feel trapped in its strangling hold.

The night before opening statements, Zoe worked late into the night at the station, triple-checking every document to ensure they were in order. As she gathered her things to leave, heavy footsteps sounded down the empty hallway. She turned to find two armed men blocking the exit. Before she could react, a cloth soaked in chloroform was pressed to her mouth. She faded into unconsciousness as her captors hoisted her limp form away.

Zoe awoke, disoriented, in a dark cellar. Shortly after her regaining consciousness, a video message, played on a loop, was sent to Paul's email. Diaz's men held a semiconscious Zoe at gunpoint. The demand was simple:

Diaz's immediate release and extraction, or Zoe would die. Word spread fast among law enforcement. Panic set in as the full gravity of the situation hit, if Diaz escaped justice now, all their sacrifices would be for nothing. Roadblocks went up as manhunts began, but the trial clock was ticking.

Paul rallied himself, determined not to give in to fear or fury. He had to believe Zoe would make it out alive, and he would be the one to finally put Diaz away for good. But it was getting late, and their enemies held all the cards. Working frantically through the night, Paul pored over case files and street reports, looking for any clue to Zoe's location. A scrap of discarded tape in an evidence photo sparked a hunch, and he raced to chase it down. It led him to another abandoned warehouse on the outskirts of town.

As dawn broke, Paul crept through shadows, senses attuned for any sign of life within. Muffled voices echoed down; he was close. Slowly ascending a rickety staircase, Paul peered through a crack in the wall. Zoe was bound and gagged, Diaz's men playing cards nearby. It was now or never for a stealthy rescue.

Paul searched his surroundings for anything that could create a distraction. Spotting a tinder-dry support beam, he lit a match and tossed it, backing away as orange bloomed. Chaos erupted as flames caught and smoke billowed. In the panic, Paul slipped behind the men, making quick work of them with precise blows. He cut Zoe's bonds just as armed backup emerged through the growing inferno. Justice would be won, one way or another, that day. Flames continued swallowing the warehouse as Paul led Zoe through the smoke.

Goons emerged from the haze, guns raised. Paul barely shoved Zoe aside as bullets whizzed by. He charged the first assailant, grappling his weapon aside and delivering a knife-hand strike. Spinning, Paul kicked another thug square in the chest, buying precious seconds. Zoe joined the fray, disarming one man with trained precision while Paul tangled with two more. Steel rang on steel as blade met blade. Through the scorching heat, Paul could see a retreating form.

A ruthless fighter blocked his path, malice gleaming in his eyes reddened by smoke. They went toe to toe in a flurry of brutal blows, the crackling inferno their backdrop with a crack of bone; Paul at last felled the man. Emerging on the landing, Paul spotted the man dragging a dazed Zoe along as a human shield. Enraged, he launched himself between burning support beams, hell-bent on bringing the criminal to justice once and for all. The climax was coming. With the smoke and flames behind them, Paul closed the distance. The killer's grip tightened around Zoe's throat as he backed toward the edge of the roof.

"Let her go," Paul demanded, slowing his approach. The killer sneered. "You've cost me everything. I'll take her with me!" As his finger curled around the trigger, Zoe dropped, sweeping the killer's legs. They struggled amid choking coughs.

Below, police swarmed into the alley. Paul caught movement behind two henchmen taken down by smoke inhalation. His eyes met Zoe's across the grappling figures, and unspoken coordination took over. She disarmed the killer at the same moment Paul kicked him away from the ledge. Cuffs clicked around flailing wrists as the criminal raged impotently at his long-coming fate.

Simon Ames, the disgruntled investor and mastermind, was finally caught. Paul was not as surprised as Zoe when they learned Ames had been pulling the strings. At last, the nightmare was over. Paul helped Zoe stand, both alive through strength, skill, and the bond of partnership that had carried them through every trial. Justice was won.

The courtroom buzzed with anticipation as Ames was brought in, cuffed but smug. He'd fired his lawyers, choosing to represent himself after Diaz surprisingly pleaded guilty. The revelation of Diaz pleading guilty to all counts was indeed surprising to Paul and Zoe. While Paul expected Diaz to bear some responsibility for the crimes committed, the fact that he admitted guilt to all charges without contest was unexpected. Throughout their investigation, Diaz had maintained a shroud of secrecy and evasion, making it challenging to gather concrete evidence against him.

The sudden turn of events, with Diaz willingly confessing to his involvement, caught Paul off guard. It was a twist in the case that forced him to reassess his perceptions and underlying assumptions. However, as a seasoned detective, Paul had learned to expect the unexpected since criminals often have their own motivations and reasons for choosing certain courses of action. While Diaz's guilty plea may seem surprising on the surface, there may be underlying factors at play that motivated his decision. The guilty plea now allows the prosecution to focus on the consequences and ensuring that justice is appropriately served to Ames. They can now turn their attention towards understanding the full extent of Diaz's and Ames' involvement and the complexities of the crimes committed, bringing closure to the case for the victims and their families.

The state readied their charges against Ames, but Paul knew this was no open-and-shut case now. This manipulator thought himself above the law, cunning enough to twist logic on a whim. When given the floor, Ames launched slickly into doubt, sowing rhetoric. Technicalities were exaggerated, and conjectured conspiracies tore apart witnesses. The jury looked wary, swaying in his grasp. Paul watched closely, waiting for his time in the witness chair. Piece by piece, he dismantled Ames' distortions, reweaving the stark truth. Witnesses regained confidence by recounting their ordeals.

By day's end, the killer's veneer showed cracks from Paul's dogged persistence. He knew the coming days would test both their resolve, but wasn't it always the darkest before dawn? Justice would prevail if he held fast to the light of truth. As the trial continued, the game was afoot. And this time, Paul was playing to win. Over subsequent days of testimony, witnesses were guided by the DA with patience and care, with Paul's help. Piece by steady piece, the DA wove together an airtight case as Ames' contorted reasoning crumbled.

Bank records exposed laundered money trails as Paul interpreted flows and shell companies. Technical analysts debunked alibis, visualizing cell data that tracked the killer to the scene. Pathologists affirmed signs of pre-death torment on Thompson's preserved remains. Slowly, Paul painted a portrait of acute calculation and sadism that had the jury transfixed. Ames struggled to poke logical holes, growing increasingly erratic. His snide remarks earned stern warnings from the judge.

By the week's end, derision had warped into uncontrolled fury. As the final medical examiner implicated the murder weapon, the criminal lunged across the defense table with an enraged howl. Bailiffs wrestled him down amid a chaos of screaming. With the order restored, the trial neared its close. Paul had done his part in presenting layered, irrefutable truth. Now, fate rested with the jury and their collective conscience. Justice would be done, one verdict or another. Paul believed deeply in the justice he helped uncover.

As the verdict was read, 'guilty on all counts,' the weight of defeat crushed Ames. Eyes burning with loathing, he muttered vile oaths until guards hauled him away. Paul watched the man's downfall, feeling only a solemn sense of justice served after so much darkness confronted him.

He turned to see Zoe smiling proudly by his side. "It's finally over," she said.

Paul nodded. Though this criminal's schemes ended here, more work remained to dismantle the rot he left in his wake. But for now, the sheer relief of prevailing overwhelmed any lingering doubts. As sirens announced the killer's transport to maximum security, Paul sensed the sinister threat receding like a shadow lifting at dawn. A new day was coming, and he could face it, knowing the lives lost had not been in vain.

Zoe took his hand. "Come on, I'll take you home." Her warmth and friendship reaffirmed Paul's belief that wherever evil existed, good people would stand against it, and light would overcome the dark.

CHAPTER 15

TWISTED FATES

Paul walked along the quiet beach, feeling the warm sand between his toes as the gentle waves lapped at the shore. For the first time in as long as he could remember, his mind was calm and free of shadows. All the uncertainty, pain, and regret that had dogged his steps for so long had finally been laid to rest. He breathed in the salty sea air deeply, savoring the sense of lightness in his soul. For years, this place, beautiful as it was, had only hinted at an escape from his past that continually evaded his grasp. But now, those demons no longer had power over him. He had faced them head-on with Zoe by his side and banished them to the darkness, where they belonged.

Looking out at the gleaming ocean stretching to the horizon, Paul felt only gratitude and relief. The cases were solved, and justice was served. He had defeated his ghosts and brought closure to the victims who had suffered for so long in the shadows. For the first time, he felt free, unburdened by failure, and able to embrace this new chapter of his life. The future, once a looming unknown, now held only promise and possibility. Paul Phillips was finally at peace.

As the sun dipped below the horizon in a blaze of color, Paul watched the waves sweep the shore in gentle, rolling swells. His thoughts lingered fondly on the past weeks and the partner who had helped him confront so many ghosts. Zoe had been a force of nature from the start, determined, passionate, and ready to battle any darkness. With her grit, intuition, and sometimes recklessness, they had succeeded where others had failed for so long. Paul knew he owed her a debt that could never fully be repaid. Yet, as the case wound down to its conclusion, Paul knew their partnership must come to an end as well. He had found the peace he'd sought in this place, while Zoe still had so much left to learn and accomplish. Though his heart was glad for all she would achieve, a part of him lamented their impending farewell.

As the stars began to dust the evening sky, Paul heard footsteps on the sand behind him. He turned to see Zoe's silhouette backlit by the fading glow of sunset. Some bonds, it seemed, were not so easily severed even with duty done. There were goodbyes still to be said and gratitude to be expressed one final time before they each walked their separate roads once more under the night sky.

Zoe had told Paul the day after the verdict that she would be leaving St. Anne. Saying goodbye to someone who had become a mentor and a friend during their time working together, Zoe was saddened by her decision to leave. She had accepted a job offer on the larger Caribbean Island of Guadeloupe. Paul had imparted valuable knowledge and guidance, and she had learned so much from him.

But along with the sadness, there was also a sense of accomplishment and gratitude. They had solved the Thompson case together, overcoming numerous hurdles and dangers along the way. Their partnership had grown stronger, and Zoe had proven herself to be a capable detective. There was a feeling of pride and satisfaction in what they had achieved. Overall, it was a complex bundle of emotions: farewell, gratitude, and a touch of sadness as she embarked on the next chapter of her career.

All too soon, the morning arrived for Zoe to depart. She and Paul lingered in the airport parking lot, delaying the inevitable goodbye. "It doesn't have to end here, you know?" Zoe said quietly. "There will always be work to do, justice left undone. I could use a partner that I trust, watching my back."

Paul smiled sadly. "And I'm honored you would ask. But this island... this is where my peace resides now. All I ever wanted was to leave that darkness behind."

"I understand. Just know the offer stands whenever you change your mind." She moved as if to turn away, then paused, looking up at him with pleading eyes. "Paul... are you certain this is what you want? Your skill could do so much good. I don't want to lose my mentor."

Paul laid a gentle hand on her shoulder, "You'll never lose me, Zoe. I'll be with you here." He touched her chest over her heart.

Reluctantly, she nodded, accepting his decision, even though it pained her. In a swift motion, she embraced him tightly, memorizing the feel of his steadiness one last time before letting go.

"Thank you for everything. The world's in good hands with you in it." Paul said.

Zoe smiled through her sadness. "Take care, old friend. And thank you." With that, she turned and boarded her flight, heading toward her next challenge while Paul watched the dawn of his new quiet life begin.

As Zoe's plane taxied down the runway, Paul felt a swell of pride in how far she had come under his guidance. Where once was a bright but overeager rookie now stood a compassionate investigator wise beyond her years. He knew with certainty that Zoe would excel at untangling life's knotty mysteries, confronting corruption with her trademark grit, and championing the voiceless. She possessed a unique ability to see into the human soul that would serve her and her community well for decades to come.

Though Paul's own journey in law enforcement had ended in shadow, he took solace in knowing their work together had helped redeem its purpose for him. Through Zoe's fearless pursuit of justice, that mantle was left in caring hands for the future.

As her plane lifted into the azure sky of the Caribbean, Paul whispered a prayer that wherever fate led her, Zoe would stay true to her gifts of empathy, tenacity, and hope. With those as her compass, no darkness could waylay her for long. With his mission now complete, Paul was ready at last to surrender the torch. He turned from the airfield toward a new dawn of peace, confident that he had played his role in passing light onto the next generation of truth-seekers. Zoe would carry it brilliantly from here.

Watching Zoe's flight depart into the sky, Paul stood gazing after it for a long while, reminiscing on their journey through darkness into light. They had walked side by side into the depths, driven by a need for truth stronger than any fear. There, in the shadows, they had found not only answers but also insight into each other's souls. An unshakeable bond was forged through facing down terrors together and emerging on the other side, liberated.

Now, as their paths diverged once more, Paul took comfort in knowing that wherever the road led from here, a part of Zoe would remain woven into the tapestry of his being. Her spirit of resilience, her fire, and her empathy would always echo in his memory.

In Zoe, too, a bit of Paul's patience, grit, and wisdom would linger, as well as armor against future storms. They had revived something noble in one another that no shadow could extinguish again. Their fates, entwined through turmoil, bid each other farewell, evolved, weapons honed, yet hearts holding fast to the light. Though separate in body, their alliance of soul would transcend all endings. This, Paul knew, was but the beginning. With that comfort, he turned to greet the new chapter in his life and the endless beginnings that lay beyond.

Along the golden shore, Paul walked, listening to the sea's gentle whisper. In the fading light, memories of Zoe rose unbidden like flickers in the deepening glow. He recalled her irrepressible grin the moment their eyes first met a resolution, her tenacity that could move mountains, and the warmth of her hand squeezing his arm in solidarity. Through each victory won together against the shadows, their bond was welded strong as steel.

Paul let the recollections wash through him, bittersweet as the breaking waves. Though the darkness was banished and justice served, a familiar ache lingered in his heart, and now their partnership was done. He knew well the pain of camaraderie lost, though also its transmutation into something more profound across the veil of time and distance. As twilight softened the horizon, Paul breathed deep the salt-sweet air laden with the spirits of allies' past.

Though worlds apart now in body, he knew Zoe would remain tethered to his soul, upholding their oath to stand as light against the encroaching dark. Here, in this place of refuge, his memory of Zoe would stand eternal guard.

As days melded into weeks of tranquil routine, Paul found his thoughts often returning to Zoe and their shared battle against the darkness. Her indomitable spirit in the face of terror was a reminder of hope's power to overcome even the deepest shadows. More than ever, Paul felt called to immortalize those victories and hard-won lessons by chronicling their journey.

Sitting on the veranda with sea and sky unfolding vast before him, he began to write. At first, Paul poured forth only memories of their recent case. But as his pen flowed, so too did recollections of decades past come flooding back: comrades fought and fallen beside, victories and mistakes that shaped who he became. Before long, what began as a memoir of one adventure transformed into a tome encompassing Paul's lifetime of service, struggle, and redemption. In sharing the intricate dance with darkness, he hoped to honor all who fought the noble battle with him through the years.

Most of all, Paul wished to offer those still waging war a glimmer of comfort, proof that even when the night seems eternal, the dawn will surely come again if one but carries on with courage, tenacity, and light. His story would be a testament to hope's inevitable triumph over even the deepest shadows of the human soul. And so, Paul wrote, and the words set free more than just memories; they released the final shackles of his past and redeemed its meaning for the future.

Months had passed since Zoe said farewell to Paul on the beach, embarking on her career with his steady light to guide her way. Now, she found herself handling her most complex case yet, a twisting conspiracy that spread throughout the city. As she paced her office late into the night, piecing together loose threads, Zoe thought of Paul's patience and felt his calm assurance warming her.

She took a breath and considered the subtleties she may have missed, seeing the bigger picture as he always advised. A new lead fell into place, and Zoe raced to the crime scene with renewed purpose. There, a clue guided her to a suspect's studio, where turmoil and clues awaited. Zoe sized up the scene with prudent care, recalling Paul's instincts she had possessed. Steadfast but cunning, she interviewed the suspect, reading between deceptions with penetrating insight. A confession emerged, confirming her deduction in a burst of triumph. Justice was served once more through her skill and will to see the entire truth unveiled.

Later, as Zoe closed her latest case file with a contented sigh, she admired the Caribbean horizon through her window. There, beyond the glimmering city lights, she sensed Paul's approving smile and grasped ever more dearly why she had chosen this life of fierce protection against the darkness. Her mentor's steadfast spirit would ensure her way for many victories to come.

As time passed, their days of partnership were memories; the bond Paul and Zoe forged through facing darkness together could never fully fade. Each carried a piece of the other wherever life led: Paul with his wisdom, Zoe with her tenacity. And while new chapters unfolded, they never forgot how confronting shadows side by side redeemed not just the victims but themselves.

For Paul, it meant finding refuge and purpose beyond what he thought possible. Zoe gained clarity, strength, and ideals to define her career in justice. Together, they overcame that which once defined them and, in victory, discovered a new purpose. Though worlds apart, whenever doubts assailed, they recalled that unyielding alliance. It remained a beacon whenever the light seemed poised to slip away again. Through it all, their fates remained luminously intertwined, a lesson that even past tragedies need not have the final say.

On a blissful morning, Paul stood beneath cloudless skies, surveying the dawn unfolding its vast canvas. All around him, nature's music swelled in joyous symphony to greet the new day. Here, in this place of solace, time had lost its power to bind. Paul was free from ghosts of past regrets or what-ifs, liberated to embrace the present unfolding moment by sunlit moment. His memoir chronicling life's twists and turns was nearly complete. Through its pages flowed absolution for paths long traveled and hope that dark truths shared may hearten others still wrestling relentless shadows. Paul's purpose was redeemed. As the sun crested gloriously, painting sea and sky in shimmers of molten gold, Paul breathed deep into the scene of paradise.

In his soul bloomed serenity to match the luminous

beauty surrounding him, for he had overcome through resilience and stood now in sunlight's blissful glow. Everything that once haunted him now nourished the man he become: guide, warrior, and friend. The future held only the promise of sunrises yet unseen. With Zoe's courage in his heart and dawn's light before him, Paul greeted the unfolding day and mysteries yet to come, finally free. With the past now behind him, he turned toward life and its endless tomorrows, at peace. This was enough. Always, always, this would be enough.